BASIC INSIGHTS IN VECTOR CALCULUS

With a Supplement on Mathematical Understanding

Other World Scientific Titles by the Author

Invitation to Generalized Empirical Method:
In Philosophy and Science
ISBN: 978-981-3208-43-8

The (Pre-)Dawning of Functional Specialization in Physics
ISBN: 978-981-3209-09-1

BASIC INSIGHTS IN VECTOR CALCULUS

With a Supplement on Mathematical Understanding

Terrance Quinn

Middle Tennessee State University, USA

Zine Boudhraa

Montgomery College, Maryland, USA

Sanjay Rai

Montgomery College, Maryland, USA

World Scientific

NEW JERSEY · LONDON · SINGAPORE · BEIJING · SHANGHAI · HONG KONG · TAIPEI · CHENNAI · TOKYO

Published by

World Scientific Publishing Co. Pte. Ltd.

5 Toh Tuck Link, Singapore 596224

USA office: 27 Warren Street, Suite 401-402, Hackensack, NJ 07601

UK office: 57 Shelton Street, Covent Garden, London WC2H 9HE

Library of Congress Control Number: 2020030528

British Library Cataloguing-in-Publication Data
A catalogue record for this book is available from the British Library.

BASIC INSIGHTS IN VECTOR CALCULUS
With a Supplement on Mathematical Understanding

ISBN 978-981-122-256-6 (hardcover)
ISBN 978-981-122-257-3 (ebook for institutions)
ISBN 978-981-122-258-0 (ebook for individuals)

For any available supplementary material, please visit
https://www.worldscientific.com/worldscibooks/10.1142/11892#t=suppl

Desk Editor: Liu Yumeng

Terrance Quinn: In memory of my parents, George and Bernice, and two of my brothers, Patrick and John, all of whom encouraged me in my interest in mathematics and science, from an early age.

Zineddine Boudhraa: In memory of my mother, Habiba, as well as the memory of all of those who lost their battles to cancer.

Sanjay Rai: To my parents, Shri Sarva Deo Rai, my late father, and Smt. Munni Rai, my mother, neither of whom had the opportunity for education, but through their own experiences in life, understood the need and potential of education in changing the lives of a generation.

We also dedicate this book to students and teachers, in the hope that it will be enjoyable and helpful in your growth in mathematics.

Preface

This book was developed from a series of lecture notes for teachers of undergraduate vector calculus, and students who want to understand the essentials of the classical theorems. With that said, we believe that graduate students in mathematics and mathematical sciences generally will also find the material helpful. In order to reach some level of mastery of, say, differential forms, one needs to have source insights for the classical vector calculus theorems. Promoting those source insights is the main purpose of our book. To help set the stage, we begin by providing some context.

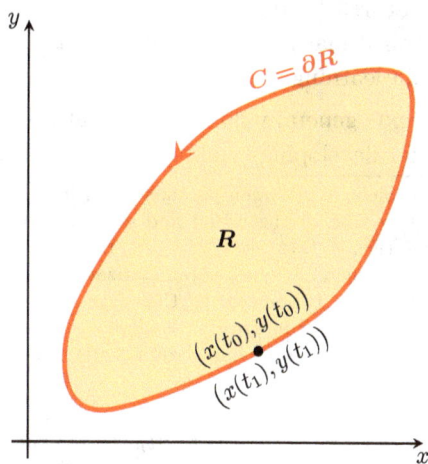

Figure P.1   Positively-oriented closed curve C.

P.1   General context

The vector calculus theorems emerged gradually through a cross-fertilization of physics and mathematics. In each of the theorems, an integral of a vector quantity along a "boundary" is found to be equal to a combination of derivatives in the region or volume "interior" to the "boundary." For instance, Green's theorem is for a vector field

$$\mathbf{v}(x, y) = \big(a(x, y), b(x, y)\big)$$

defined on a region R in the (x, y) plane. The boundary of R is a "positively-oriented" closed curve[1] C (see Figure P.1). With appropriate hypotheses in place,

$$\oint_C a\, dx + b\, dy = \iint_R (b_x - a_y)\, dx dy = \iint_R \mathrm{curl}(a, b)\, dx dy\;,$$

where

$$\mathrm{curl}\big(a(x, y), b(x, y)\big) := b_x(x, y) - a_y(x, y).$$

The vector calculus theorems are standard content for undergraduate multivariable calculus books presently on the market.[2] There is, however, a gap in the textbook literature. Filling that gap will meet a need for students in both pure and applied mathematics. Note that the gap is not in topics treated, but in certain learning outcomes.

Currently available texts generally do a good job at providing applications, as well as examples for developing symbolic and computational techniques.

---

[1]In this context, closed means that when parameterized in a permissible way (for instance, one-one and differentiable) the initial and final points of $\mathbf{r}(t) = (x(t), y(t))$, $t_0 \le t \le t_1$ are the same, that is, $(x(t_0), y(t_0)) = (x(t_1), y(t_1))$. For the term *positively-oriented*, non-technically, one can imagine a parameterization tracking counter-clockwise relative to a point in the region R bounded by the curve. The "circle" on the integral symbol indicates that the curve for integration is closed. To compute the *line-integral* $\oint_C a\, dx + b\, dy$, we use any permissible parameterization $\mathbf{r}(t)$ to get

$$\int_{t=t_0}^{t=t_1} \left[a\left(x(t), y(t)\right) \frac{dx}{dt} + b\left(x(t), y(t)\right) \frac{dy}{dt} \right] dt.$$

The integral of $\mathrm{curl}(a, b)$ is the familiar 2-d integral from 2-variable integral calculus. In order for the formulas to be well-defined, it is crucial that the integrals involved be invariant under permissible changes of variables. That follows from the chain rule in finite dimensions. Details are brought out in the book.

[2]One may explore the literature.

And in most books, there are proofs — or at least partial proofs — of the vector calculus theorems.

In modern mathematics, proofs are needed. As you will find by adverting to your own experience, however, prior to proof, there is *basic insight*. For mathematicians and mathematics students, *basic insight* is a familiar event.[3] It is an insight, through a diagram or symbolism, or some combination in one's senses. Basic insight is not a guess, but this is not to suggest that guessing does not occur. Basic insight, however, is something more. It sometimes comes only after considerable effort, such as a breakthrough to what might turn out to be a key result in a Ph.D. thesis. Or, one may discover a possible solution to a problem in a homework set. The emergence is not routine. In basic insight, we grasp possibility.

If, however, one goes on to ask 'Is it so?', then one has shifted to a new kind or "level" of inquiry. Again, it is a matter of adverting to your own experience. Asking 'Is it so?' is asking about something that, to some extent, one already understands, at least as a possibility. Is the "it" that one has grasped correct? One has had an insight. It is not nothing. But does one's understanding hold up? Are there counter-examples that might invalidate one's conjecture, or cases not accounted for? Is correction or revision needed? Is what one understands partly mathematical but also partly descriptive of diagrams, such as when "outward unit normal" is only descriptively defined in coordinate geometry?[4] And so on. All of this is familiar to the mathematics student and the professional. In modern mathematics, asking "Is it so?" calls for *proof*. That is, one seeks "middle terms"[5] by which that which one has discovered as a possibility is understood to cohere with other results that have already been understood to belong to an (axiomatic) context.

Basic insight, then, is crucial. It is a beginning. It emerges from a 'What is it?' inquiry-mode, wherein one grasps possibilities. Being asked to read a proof prior to being invited to basic insight is being asked to read a solution to a problem that has not yet had. And, without basic insight, the possibility of competence with symbolic and computational techniques is significantly undermined.

---

[3] For more detailed observations, see the Supplement.

[4] Rigorous definition of *orientation* requires a more advanced context. See Supplement, note 31, for an indication of the importance of identifying "casual insights."

[5] See Supplement, Section A.4, for more details on "proof" and "middle terms."

In calculus textbooks presently available, however, the emphasis tends to be on symbolic technique and computation.[6] This Preface is not the place for a literature review. To help illustrate the problem, it will be enough to advert to two common ways that "curl" is introduced.

P.2   Beginning with the coordinate definition of the 3-d vector curl

Frequently, the discussion of "curl" begins by providing a formula for the 3-d "vector curl" for $\mathbf{F} = P\mathbf{i} + Q\mathbf{j} + R\mathbf{k}$. Assuming that all partial derivatives exist,

$$\operatorname{curl} \mathbf{F} := \left(\frac{\partial R}{\partial y} - \frac{\partial Q}{\partial z} \right) \mathbf{i} + \left(\frac{\partial P}{\partial z} - \frac{\partial R}{\partial x} \right) \mathbf{j} + \left(\frac{\partial Q}{\partial x} - \frac{\partial P}{\partial y} \right) \mathbf{k}.$$

This is then followed by discussions suggesting that "curl" represents rotations in, for example, a fluid flow along a surface. And so, when $\operatorname{curl} \mathbf{F} = \mathbf{0}$, a fluid is said to be "irrotational." Examples follow where the vector curl is computed for known vector fields. Justification for the association with fluid flow is then provided, ultimately, by appealing to Stokes' Theorem in three dimensions. Examples can provide plausibility. Note, however, that Stokes' Theorem is a generalization of Green's theorem in the plane to 2-d surfaces in (x, y, z) space. Appealing to the generalization directs student attention still further away from the possibility of basic insight. What is the basic insight that initially leads to the definition of the 3-d (or 2-d) coordinate definition of curl?

P.3   Beginning with descriptions of imagined paddle wheels and integral definitions of *circulation*

A different but common approach is to begin by describing merely imagined fluid flow and then inviting the reader to think of how a paddle wheel might rotate, or not, when immersed in the fluid flow at various angles.

Following thought experiments about water flow and paddle wheels, a "circulation density vector" in the direction of a normal vector $\mathbf{n}$ is defined

---

[6]This includes "calculus reform." See Sections P.2 and P.3. The books by Snider and Davis (2000) and Lovrić (2007) both include some pedagogical strategies that are "on point." See Section P.6.

by:

$$\lim_{\text{Area}\to 0} \left[\frac{\text{Circulation around } C}{\text{Area inside } C} \right] = \lim_{\text{Area}\to 0} \left[\frac{\int_C \mathbf{F} \cdot d\mathbf{r}}{\text{Area inside } C} \right].$$

Note, however, that what happens to paddle wheels in water flow is only settled by experiment. As it happens, if a handle with a paddle wheel is placed horizontally, then in some cases the paddle wheel rotates, while in other cases it does not. Whether or not a paddle wheel rotates also depends on orientation of the handle relative to the direction of flow. The integral definition given above comes from classical fluid dynamics and, with some work, can lead to the 3-d vector curl formula given in Section P.2. Discussion of "imagined push and pull of water flow" notwithstanding, without computing the link from the integral to the vector form, the student is directed away from the possibility of basic insight and here, too, is instead moved along into a flow of exercises for symbolic techniques.

While descriptions of water flow can invite beginnings, it took physics decades to formulate key quantities and vector formulas. There are sequences of insights needed, to go from imagining and describing water motion to understanding such motion in terms of vector fields, vectors normal to a plane in which "rotational displacement" occurs, combinations of partial derivatives and line integrals.

P.4 Do vector fields represent force or velocity? Or other?

In undergraduate multi-variable calculus books, *line integrals* are often motivated by appealing to the definition of *work*, from classical physics. In simple cases, Work = (Force) × (Distance). More generally, the work done by a force vector field $\mathbf{F}$ along a curve C is defined to be the line integral $\int_C \mathbf{F} \cdot d\mathbf{r}$.

On the other hand, *flux* and *circulation* integrals are usually motivated by asking the reader to think of the vector field $\mathbf{F}$ as representing a *velocity* field. Then, with no further advisories, formulas for velocity fields and for force fields are combined in the same vector calculus theorems.

The vector calculus theorems are, of course, correct. This approach, however, covers over fundamental insights and, for the unwary, introduces various errors. In both classical physics and calculus, flows, velocities and forces are not directly comparable. When mass is constant, force is the derivative of $(mass) \times (velocity)$. In what way, then, do examples computing line integrals of force vector fields $((mass) \times (acceleration))$ help a beginner understand *circulation* when in fluid dynamics *circulation* is something else? In fluid dynamics, *circulation* is an *average mass-flow rate* obtained from a *velocity* field. As in Section P.3, *circulation density* is a subtly defined *circulation per unit area*.

P.5   Historical context and pedagogy

In multi-variable calculus books, the approach for teaching "curl" tends to be "too mathematical." Why do we say that when the theorems in question are the fundamental mathematical theorems of vector calculus? It is because initial inquiry and understanding that led (and leads) to their discovery (e.g., the definition of "curl" in 2 dimensions) are about fluid motion and other field quantities (electricity and magnetism). Indeed, the vector calculus theorems were discovered by mathematical physicists.[7] And not adverting to that obfuscates main sources of understanding.

This raises questions about undergraduate curriculum structures. Nowadays, students who take undergraduate courses in Calculus III or Vector Calculus are from majors including, but not limited to, mathematics, engineering, aerospace, geography, oceanography, mathematical statistics, biology, and mathematics education. It would not be practical to ask that all students from all majors who are wanting a working knowledge of vector calculus also take courses in mathematical physics. This raises questions about questions. What are questions and key insights needed to reach basic understanding of the fundamental theorems of vector calculus? And, are they accessible to students from a variety of majors?

This is not a "speculative" question. It is about inquiry and insights that are already present in the tradition.

What we did, then, was sift through the historical development of classical mathematical physics, with an eye on fluid dynamics and the vector theorems. This included re-reading some of the older undergraduate textbooks for physics students, with titles such as "Introduction to Mathematical Physics." There is no new mathematics or physics in this book. Nor are

---

[7]Regarding the historical origins of "Green's theorem," Katz observed that "[a]ll of the mathematicians who stated and proved versions of this theorem were interested in it for specific physical reasons" (Victor J. Katz, "The History of Stokes' Theorem," *Mathematics Magazine*, vol. 52, no. 3 (May, 1979), pp. 146–156. See p. 149.)

we trying to teach mathematics based on "some model of how the topic in question can be learned."[8] We are, instead, drawing on what actually has gone forward in classical mathematics and mathematical physics.

Note, however, that we are not attempting to implement the "historical method of teaching" that has been popular in some quarters in the last 20 years or so. This is not a reflection on that method. We are guiding students toward particular insights that are part of the tradition, even if some of those insights have not yet been adverted to and are found in diverse sources. What is new, then, is not only an identification of insights, but an identification of "genetic sequences" of accessible and, as it turns out, elementary insights. And so we have found that undergraduates with only modest background have sufficient background to climb into the context of the fundamental theorems of vector calculus. The lessons of the book are organized accordingly.

For that reason, we do not start with definitions of vector fields or their derivatives, nor with line integrals. Partly taking our lead from early textbooks in mathematical physics, the lessons begin by inviting attention to what can be described, namely, fluids in motion as well as test particles in fluids in motion. (Most students will not have access to apparatus for studying fluid motion. But we hope that most students have an opportunity to observe water flowing somewhere.) To avoid confusion caused by talking about integrals for *work* and integrals for *circulation* in the same context, we keep the focus throughout on streamlines and first derivatives called *velocity fields*.[9] In that context, various questions arise naturally. For instance, there is the problem of understanding describable rotation in fluid flows. As in physics, we can use first-order approximation to estimate differences in locations of test-particles in different streamlines. There is no need for metaphors such as "tendency to rotate." Instead, the formula $\mathrm{curl}(a, b) = [b_x - a_y]$ is discovered as a precise solution to a focused inquiry.

---

[8]Ed Dubinksy, "Using a Theory of Learning in College Mathematics Courses," *MSOR Connections* (2000), 1.10.11120/msor.2001.01020010. This article was originally published in Newsletter 12, TaLUM, Teaching and Learning Undergraduate Mathematics subgroup, 2001. It is available online: www.math.wisc.edu/ wilson/Courses/Math903/UsingAPOS.pdf. Accessed July 17, 2019. See Supplement, Part C, for discussion.

[9]The approach taken is guided by the historical development in mathematics and mathematical physics. It is the approach taken in modern differential geometry, where streamlines and velocity fields are the basic elements. Circulation is an average weighted velocity along a curve. Force is a second derivative. Work is an average weighted force along a curve.

The background needed is two-variable calculus and an understanding of tangent vectors of streamlines.

Once the student is comfortable in understanding the vector calculus theorems for *velocity vector fields*, it is not too difficult to go back over results and realize that most of the mathematical aspects of the development do not rely on the physical significance of the vector field. In that way, then, the student gets two other sets of applications, namely: (1) in physics, for a vector field that is a *force* vector field; and (2) in mathematics, for all mathematical vector fields for which hypotheses are satisfied.

P.6   Prerequisites, the lessons, and follow up

For the student reader, formal prerequisites include: a semester each of single-variable differential and integral calculus; and some experience in multi-variable calculus. Having some understanding of the multi-variable chain rule will help. However, that result, as well as applications in integration, are reviewed in Sections 1.9–1.13. For readers who already have some linear algebra in their background, a few supplementary results are provided at the end of Part 1. These shed additional light on the equivalence of Green's Theorem and the 2-d Divergence Theorem. We do that by looking to the Jacobian (that is, the first-order approximation) of the velocity vector field. Similar results are included at the end of Part 2, in 3 dimensions. In that case, one looks to the 3×3 Jacobian.

Part 1 is for results in the (x, y) plane. Part 1 leads up to Green's theorem and (in the 2-d plane only, its equivalent) the 2-d divergence theorem. Part 2 of the book is for three dimensions. It leads up to Stokes' theorem and the 3-d divergence theorem.

The sections are brief. We have included problems and exercises.

Basic insights in vector calculus will be helpful in many areas in the undergraduate Science, Technology, Engineering and Mathematics programs. These include fluid dynamics as well as more advanced mathematics in axiomatic contexts.[10]

Follow-up material at the undergraduate level can be found in, for instance, the following two books, both of which provide far more complete coverage of topics than the present book: Harry F. Davis and Arthur David Snider, *Introduction to Vector Analysis*, 7th ed. (Mt. Pleasant SC: Hawkes Learn-

---

[10]See Katz, 1979, "Generalization and unification," pp. 150–156.

ing, 2000); and Miroslav Lovrić, *Vector Calculus* (Hoboken NJ: John Wiley and Sons. Inc., 2007). The book by Davis and Snider includes discussions of some of the more advanced equations of mathematical physics. See, e.g., Appendix C, The Vector Equations of Classical Mechanics (339–351) and Appendix D, The Vector Equations of Electromagnetism (352–371) and, of course, the equation of continuity (123). They provide two mathematical derivations of the 2-d curl (125–132). Readers of the present book will be in good shape to read the proof of Stokes' theorem (298) given in the book by Snider and Davis. The book by Lovrić provides some mathematical proofs for special cases. It has an especially rich supply of exercises and examples. It also includes introductory level discussion of some of the equations from classical physics. See, e.g., Fluid and Heat Flow (520–522), Vector Calculus in Electromagnetism (553–566), and Vector Calculus in Fluid Flow (566–576). The book by Lovrić also has a brief but convenient technique-oriented introduction to differential forms, Differential Forms and Classical Vector Integration Theorems (536–552).

P.7 Our experience

In American colleges and public universities, the third semester of a calculus sequence normally includes multi-variable calculus. In that third semester, many topics are included and there is usually not enough time to thoroughly treat the fundamental theorems in both two dimensions and three dimensions. When time is limited, we have found that students manage better when the focus is mainly on the 2-dimensional case, treated at a more leisurely pace and in some detail. This allows the teacher to promote basic insights needed for Green's Theorem and the 2-d Divergence Theorem. This avoids having to rush through complex vector theorems, which can leave students with a confused jumble of integral formulas. On the other hand, we have found that by first promoting basic understanding of the vector calculus theorems in 2 dimensions, a few lectures at the end of a Calculus III semester with examples and pointings to the 3-d cases are generally well-received.

In a full semester of Vector Calculus, we have found that by promoting basic insight, students from diverse majors have been able to do end-of-semester projects. In those projects, students are asked to do original work. We don't mean developing "original theorems." Rather, in an area of their choice, with each project pre-approved by the instructor, students are asked to work out details of applications of the vector calculus theorems. We have found that students tend to take happy ownership of such projects. It is delightful and inspiring to witness the wide range of presentations that reveal students' understanding, as well as ever further applications of the fundamental theorems of vector calculus.

P.8   Acknowledgments

We are indebted to colleagues of Sanjay Rai and Zine Boudhraa at Montgomery College, MD. Motivation for this book partly emerged in the context of conversations regarding possible redesign of particular elements of the undergraduate mathematics curriculum. Some of these discussions regarded the teaching of single- and multi-variable calculus, and the importance of understanding, and being in control of concepts. The divergence and curl operators especially illustrated the need for insight regarding what they convey about a vector field. We would like to thank our students who help us grow as teachers. Sanjay Rai would like to thank his professors from Allahabad University, Dalhousie University and the University of Arkansas for their tutelage and inspiration. Grateful to all of his teachers, Terrance Quinn would like to thank three, in particular, who introduced him to the vector calculus theorems, in various contexts: Professor Rooney (1925–2016), in Advanced Calculus (University of Toronto, Fall and Winter of 1981/82). Rooney opened our eyes to riches of classical mathematics; Professor Keith Johnson (Dalhousie University) helped us learn about differential forms (Spring, 1987) from Michael Spivak's deceptively brief *Calculus on Manifolds*[11]; and Professor Lowell Jones who (in a graduate course at SUNY at Stony Brook, Fall, 1988) shed light on the significance of the vector calculus theorems in 20th century differential geometry.[12] Finally, we'd like to thank two anonymous reviewers for their comments and suggestions which helped improve the book.

---

[11]Michael Spivak, *Calculus on Manifolds. A Modern Approach to Classical theorems of Advanced Calculus*. Reading, MA: Addison-Wesley, 1965.
[12]The textbook used was the classic: Noel J. Hicks, *Notes on Differential Geometry*. New York: Van Norstrand Publishing, 1965.

Contents

Chapter 1

REVIEWING SOME CALCULUS ESSENTIALS

The 2-d and 3-d vector calculus theorems are extensions of the Fundamental Theorem of Calculus in one variable that students encounter in a first course in calculus. To understand the vector calculus theorems, one needs a good understanding of the one variable case. Sections 1.1 and 1.2 review some key ideas in single variable calculus, leading up to the fundamental theorem in one variable. Section 1.3 is to help bring out why the fundamental theorem of calculus is called "fundamental." (Among other things, it led to the solution of numerous previously unsolved problems. It also opened up new lines of inquiry.) Sections 1.4 and 1.5 foreshadow how the fundamental theorem in one variable is both clue and key for the vector calculus theorems. In the 2-d and 3-d cases, the theorems involve areas and angles between vectors. Sections 1.6 and 1.7 are for developing the dot product, and the vector product in three dimensions. The dot product and the vector product are both solutions to problems in classical coordinate geometry. The cross-product rule is a technical result that will be needed. Sections 1.8, 1.9 and 1.10 discuss three main cases. Section 1.11 draws attention to implications in integration over a region in the plane.

This chapter is for development in prerequisite material. If you already have a good understanding of one variable calculus and other topics treated here, you could go directly to the beginning of Part 1, Fluid Motion in 2 Dimensions. You could then re-visit topics treated in this chapter on a "need to know basis."

1.1   Rates of change in one variable

Galileo discovered the "law of falling bodies." He began by slowing motion to "free-roll," by rolling steel balls down polished beams of wood, set at various angles to the ground.

Figure 1.1   Galileo's experiments. Rolling steel balls down wooden beams. Measure distances and times.

He had a radically new idea: compare measured distances and measured times.[13] By changing the angle and length of the beams of wood, he studied numerous cases. In tables of numerical data, he discovered that, in each case, distance was (approximately) proportional to the square of time, $s = kt^2$. The constant of proportionality k depended on angle of the beam. Using elementary geometry, Galileo extrapolated to the case of a "vertical beam"[14] and obtained his now famous result, "the law of falling bodies."

In today's units, the law of falling bodies is $s = 16t^2$ feet in t seconds.[15] By selecting an appropriate angle for the beam of wood, we treat the case $k = 1$, that is, $s = t^2$ feet in t seconds.

Recall that *average speed* is a ratio,

(increment of distance)/(corresponding increment of time).

---

[13] How did he measure "time"? Galileo describes using both a pendulum and a water clock. Discovering a correlation between measured distances and times was the beginning of modern physics. Regarding Galileo's work, there remain unanswered questions. See, e.g. Lindberg, David C. "Galileo's Experiments on Falling Bodies," *Isis* 56, no. 3 (1965): 352-54. http://www.jstor.org/stable/228111.

[14] In modern terminology, the constant k is proportional to sin(angle between beam and ground). This reaches a maximum when the angle is 90°.

[15] This is approximate, ignores wind resistance and supposes "other things being equal."

Straightforward computation shows that for $s = t^2$, average speed in the fourth second is larger than average speed in the third second, average speed in the fifth second is larger than average speed in the fourth second, and so on. In other words, free-fall is "accelerated motion."

A question arises: What is the speed of a free-falling object falling near a particular time, near, say, $t = 3$?

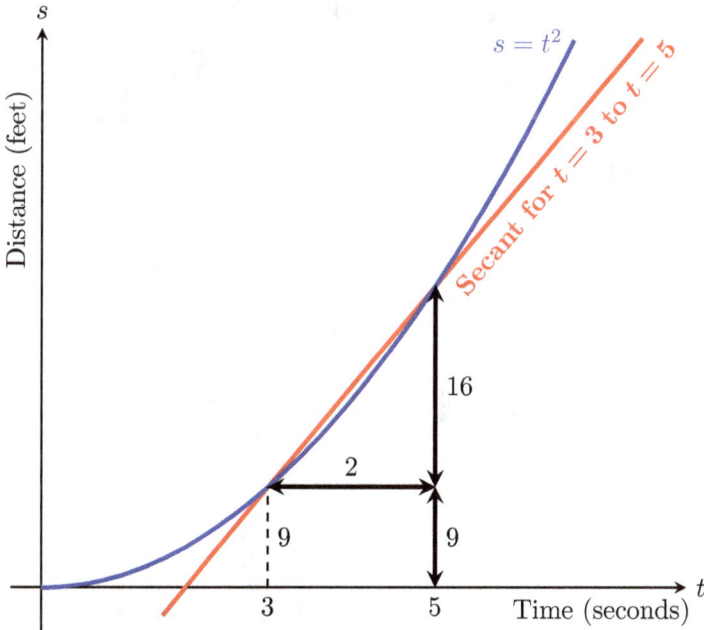

Figure 1.2 Distances, times and average speeds are conveniently represented in a graph. Average speed for a time interval is slope of a secant. Slope of secant for $t = 3$ to $t = 5$ is $\frac{25-9}{5-3} = \frac{16}{2}$.

Students who have been through a first course in calculus will be familiar with the following computations.

For $t = 3$, and with the idea of improving accuracy in computing average velocity, we use increasingly small time intervals.

In units of $\dfrac{\text{feet}}{\text{second}}$, for the time interval $t = 3$ to $t = 3+1$, average speed is

$$\frac{\text{distance}}{\text{time}} = \frac{(3+1)^2 - (3)^2}{(3+1) - (3)} = \frac{\left[3^2 + 2 \cdot 3 \cdot (1) + (1)^2\right] - (3)^2}{(1)}$$

$$= 2 \cdot 3 + (1).$$

For the time interval $t = 3$ to $t = 3 + \dfrac{1}{10}$, average speed is

$$\frac{\text{distance}}{\text{time}} = \frac{\left(3 + \dfrac{1}{10}\right)^2 - (3)^2}{\left(3 + \dfrac{1}{10}\right) - (3)} = \frac{\left[3^2 + 2 \cdot 3 \cdot \left(\dfrac{1}{10}\right) + \left(\dfrac{1}{10}\right)^2\right] - (3)^2}{\left(\dfrac{1}{10}\right)}$$

$$= 2 \cdot 3 + \left(\frac{1}{10}\right).$$

For the time interval $t = 3$ to $t = 3 + \dfrac{1}{100}$, average speed is

$$\frac{\text{distance}}{\text{time}} = \frac{\left(3 + \dfrac{1}{100}\right)^2 - (3)^2}{\left(3 + \dfrac{1}{100}\right) - (3)} = \frac{\left[3^2 + 2 \cdot 3 \cdot \left(\dfrac{1}{100}\right) + \left(\dfrac{1}{100}\right)^2\right] - (3)^2}{\left(\dfrac{1}{100}\right)}$$

$$= 2 \cdot 3 + \left(\frac{1}{100}\right).$$

And so on!

That is, for the time interval $t = 3$ to $t = 3 + h$, average speed is

$$\frac{\text{distance}}{\text{time}} = \frac{(3+h)^2 - (3)^2}{(3+h) - (3)} = \frac{\left[3^2 + 2 \cdot 3 \cdot (h) + (h)^2\right] - (3)^2}{(h)}$$

$$= [2 \cdot 3 + (h)] \,\frac{(\text{feet})}{(\text{second})}.$$

On the graph of $s = t^2$, these are slopes of secants that get increasingly close to being tangent to the graph at $t = 3$.

The algebra reveals that the smaller the time interval h, the closer $[2 \cdot 3 + (h)]$ is to $2 \cdot 3$, that is, the closer the average speed is to $2 \cdot 3 = 6$. Saying it another way, as h approaches zero, the difference between the difference ratio $\dfrac{(3+h)^2 - (3)^2}{(3+h) - (3)}$ and 6 approaches zero.

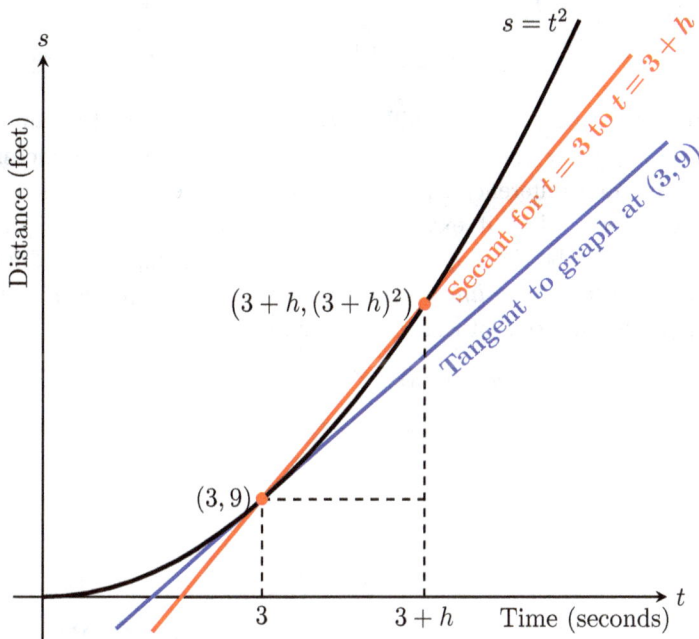

Figure 1.3 For h small, the secant is close to being the tangent to the graph at $(3, 9)$; and the slope of the secant is close to being the slope of tangent line.

The quantity $2 \cdot 3 = 6$ is called the "instantaneous" *rate of change* at $t = 3$. The adjective "instantaneous" has often led to confusion. If we restrict to time $t = 3$, there is no average speed since for average speed we need two times and two possibly different locations. The name "instantaneous" is, however, a convenient name for the entire computation and rationale given above.

By an insight, the *instantaneous rate of change* is abstracted from sets of fractions of average rates of change.

On the graph, average rates are the slopes of secants; and the smaller the time interval the closer the secant is to the being tangent to the graph. The instantaneous rate of change $2 \cdot 3$ is, then, the slope of the tangent line of the graph of $s = t^2$ at the point $(3, 9)$.

General Case: In a first Calculus course, one is asked to determine rates of change. One begins with *average rates of change*, that is, difference ratios of the form

$$\frac{f(t_0 + h) - f(t_0)}{h}.$$

When represented on a graph, difference ratios are slopes of secants. Following the approach described above, one increasingly reduces increments h. By examining difference ratios, one identifies the "instantaneous rate of change" at t_0 (when it exists, mathematically). For the graph, that quantity is said to be the slope of the tangent line at $(t_0, f(t_0))$. That tangent line is said to be the *linear approximation to the graph, near the point* $(t_0, f(t_0))$. The equation of that tangent line is

$$y = f(t_0) + f'(t_0)\,(t - t_0).$$

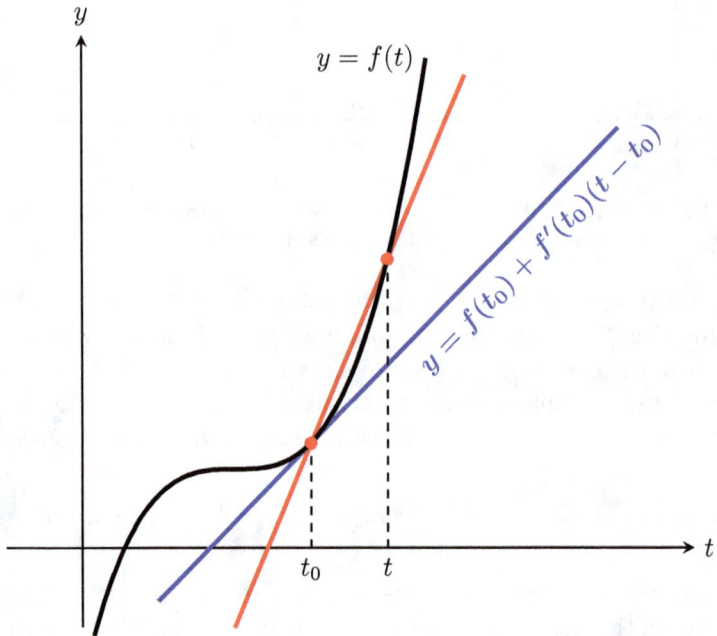

Figure 1.4   Linear approximation to $y = f(t)$ is the tangent line given by $y = f(t_0) + f'(t_0)(t - t_0)$. The time difference $t - t_0 = h$.

Notation: Using Leibniz's notation, the *instantaneous rate of change* is written $\frac{df}{dt}$.[16] Other notation that is also now standard goes back to Newton and Lagrange, a dot and a prime, respectively (convenient when, for instance, the independent variable does not require special attention). And so, while there are other notations, in calculus and its applications the following three are common:

$$\frac{df}{dt} = \dot{f}(t) = f'(t).$$

More generally, the *derived* quantity $\frac{df}{dt}$ is called the *derivative at t*. Since the derivative usually is computed for more than one value of t, $\frac{df}{dt} = \dot{f}(t) = f'(t)$ is also called the *derivative function*.

Derivatives and Anti-derivatives: Once the derivative is understood, we can study the derivative of sums, products and quotients of functions. That way we can build up a repertoire of derivative formulas. By the same token, we can work out tables of "anti-derivatives." For instance, $\frac{d}{dx}(x^2) = 2x$. If, then, we start with $g(x) = x$ and ask for the function $f(x)$ that satisfies $\frac{d}{dx}f(x) = x$, the solution is $f(x) = \frac{x^2}{2}$.

Exercises

Exercise 1.1.1
Suppose a ramp is at an angle so that Galileo's result for "free-roll" is $s = t^2$ feet in t seconds. (If you do a course in Newtonian physics, you will be able to compute the angle at which that occurs. For now, suppose that in Galileo's experiment, free-roll $s = 16\sin(\theta)\,t^2$. Find the angle for which $s = t^2$.)

(a) Compute the average speed for the time interval $t = 6$ to $t = 6.5$.
(b) Compute the average speed for the time interval $t = 6$ to $t = 6.1$.
(c) Compute the average speed for the time interval $t = 6$ to $t = 6 + h$, where h is any small positive number.
(d) Compute the "instantaneous" speed at $t = 6$? Explain.

---

[16]Using lower case d's, Leibniz's notation reminds one that the derivative is obtained from difference ratios of the form $\frac{\text{change in } f}{\text{change in } t}$.

Exercise 1.1.2

Do Exercise 1.1.1 for a more general case. That is, as is done in the book for $s = t^2$, use only "first principles" to find the "instantaneous speed" at any time t during free-roll.

Exercise 1.1.3

Suppose that a metal shuttle attached to a spring slides along a smooth surface (friction is negligible) and that, for a time, its motion approximately satisfies $s = t^3$ feet in t seconds. Compute a good estimate for the "instantaneous speed" at time t.

Exercise 1.1.4

Suppose that $f(x) = \dfrac{x^2}{2}$. Using only first principles show that

$$\frac{d}{dx} f(x) = x.$$

Exercise 1.1.5

Suppose that $f(x) = x^3$. Using only first principles show that

$$\frac{d}{dx} f(x) = 3x^2.$$

Exercise 1.1.6

Suppose that $f(x) = \cos x$ and $g(x) = \sin x$, where x is in radians. Using only first principles, show that

$$\frac{d}{dx} f(x) = -\sin x \ \text{ and } \ \frac{d}{dx} g(x) = \cos x.$$

Hint: What are the addition formulas for $\cos(x + h)$ and $\sin(x + h)$? Looking to the unit circle, for $h \approx 0$, $h \neq 0$, what are geometric approximations for $\dfrac{\cos h}{h}$ and $\dfrac{\sin h}{h}$? Remember what radians means, a unit distance. What is that unit, in the case of a circle of radius 1?

1.2   The fundamental theorem of calculus in one variable

Computations such as those in Section 1.1 are well-known. Moreover, both linear approximation and being able to find anti-derivatives are helpful in solving many problems in mathematics and applications. But these are not yet The Fundamental Theorem of Calculus.

To discover the fundamental theorem of calculus, we look to the problem of determining the instantaneous rate of change of "area."[17]

Consider, for instance, the area of a square whose length is x. The area is $A(x) = x^2$.

If we start with, say, a square of length $x = 3$, then its area is $A(3) = 3^2 = 9$. If the length of that square is increased by $dx = 1$, then the area of the new square is $A(3 + dx) = A(4) = 4^2 = 16$. The change in area is

$$dA = A(3 + dx) - A(3) = A(3 + 1) - A(3) = 7.$$

If, however, we start with a square of length $x = 10$ and again increase length by $dx = 1$, then

$$dA = A(10 + 1) - A(10) = 121 - 100 = 21.$$

The problem is to determine how area of a square changes *relative to change in length*.

The computations needed are described in Section 1.1. In that sense, numerically, the answer is already known. But the new question is not merely numerical. We ask about "rate of change of *area* relative to *change in length* of area." There is, then, a further insight to be had, an insight that (in this case) pertains to the geometry of the square.

Before going on with the discussion about square areas, however, it may help to pause over two simpler examples.

Imagine two rolls of paper. One of these is 24″ wide; and the other is 40″ wide.

Imagine the 24″ wide roll already rolled out across a flat cutting board. If it is rolled forward an additional 1″, additional area obtained is 24″ × 1″ = 24 square inches.

If we do the same computation for the 40″ wide roll and it too is rolled forward an additional 1″, then the additional area is 40″ × 1″ = 40 square inches.

What is it about the rolls of paper that provides different answers?

For the 24″ wide roll, the rate of change of area is

(24 square inches)/(inch that the roll is advanced);

and for the 40″ wide roll, the rate of change of area is

(40 square inches)/(inch that the roll is advanced).

---

[17]See note 7.

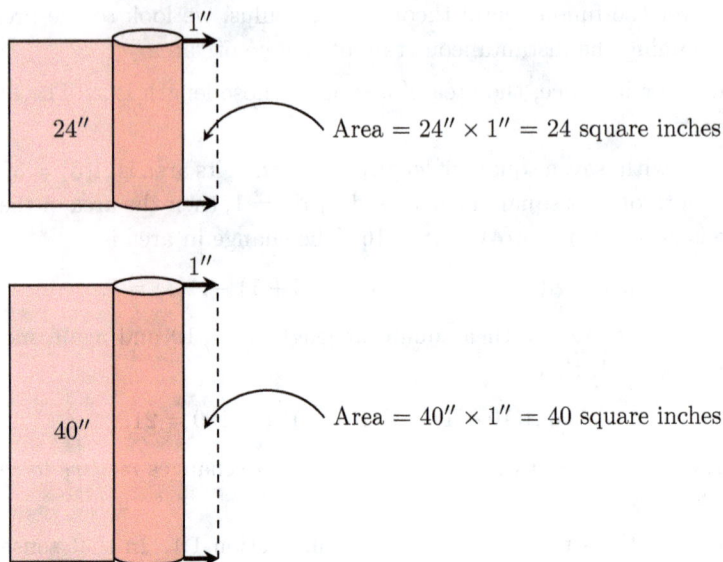

Figure 1.5   Two rolls of paper, of width 24″ and 40″ respectively. Rolling forward, the rate of change of area is (24 sq. inches)/(inch advanced) and (40 sq. inches)/(inch advanced), respectively.

In other words, for a given roll of paper, the rate of change of area is the width of the roll!

Let's now go back to the question about square areas. After some progress in differential calculus, we have that

$$\frac{d}{dx}(x^2) = 2x.$$

But the new question asks for something new:

What is the rate of change of a square area relative to change of length?

For instance, for $x = 3$ to $x = 3 + \dfrac{1}{100}$, there is an initial 3×3 square nested inside a larger

$$\left(3 + \frac{1}{100}\right) \times \left(3 + \frac{1}{100}\right)$$

square (see Figure 1.6).

In terms of the length of the initial 3×3 square, what is the ratio

$$\frac{\text{change in area}}{\text{change in length}}?$$

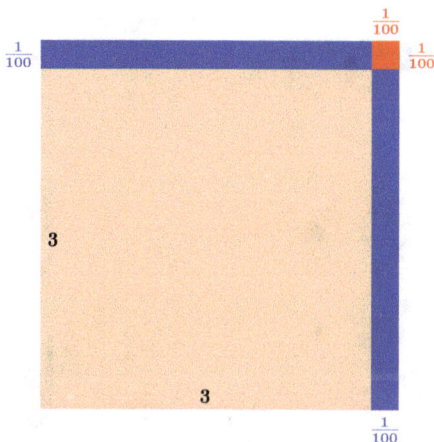

Figure 1.6 [Area of square of length $(3 + \frac{1}{100})$] - [Area of square of length 3] = [Areas of two rectangles] + [Area of corner piece].

By appealing to the diagram, the change in area consists of **two** identical rectangles together with a corner piece. Each rectangle is of length 3 and width 1/100. To get the average rate of change, we need to divide the sum of these "leading edge areas" by the change in length that produces the new square.

That is,

$$\frac{\text{change in area}}{\text{change in length}} = \frac{\mathbf{2} \cdot 3 \cdot \left(\frac{1}{100}\right) + \left(\frac{1}{100}\right) \cdot \left(\frac{1}{100}\right)}{\left(\frac{1}{100}\right)} = \mathbf{2} \cdot 3 + \left(\frac{1}{100}\right).$$

The general case is similar:

Let h be a small change in length of any square of length x.

Then, again by appealing to the diagram,

$$\frac{\text{change in area}}{\text{change in length}} = \frac{\mathbf{2} \cdot 3 \cdot (h) + (h) \cdot (h)}{(h)} = \mathbf{2} \cdot 3 + (h).$$

Do you see, in terms of the geometry, why there is a factor of 2?

By increasing the length of a square to produce a new square, the initial square is expanded along **two** front-lines! The result is essentially the same as for the rolls of paper discussed in the examples above.

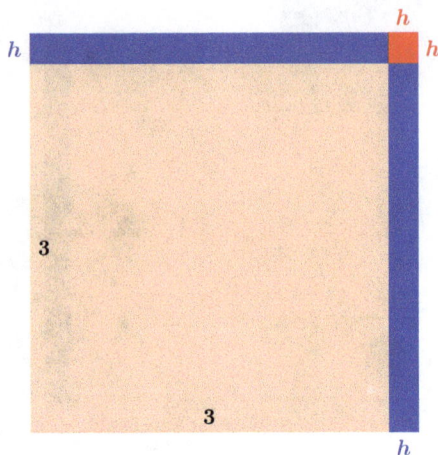

Figure 1.7   [Area of square of length $(3+h)$] – [Area of square of length 3] = [Areas of two rectangles] + [Area of corner piece].

That is:

Relative to change in length x, the instantaneous rate of change of a square area $A(x) = x^2$ is the length of the advancing front line. In the case of a square of length x, that length happens to be $2x$.

Without yet having a rigorous definition of *area*,[18] the same question can be asked about an area $A(x)$ between the graph of a non-negative function $y = f(x) \geq 0$ and the x axis.

Notation: The symbolism for 'area under a graph' was introduced by Leibniz (1646–1716):

For the interval $a \leq t \leq x$, the area $A(x)$ is written

$$A(x) = \int_a^x f(t)dt,$$

where t is a "dummy" variable.[19] The Leibniz notation is to express the idea that the total area under a graph is somehow a sum of thin slices of

---

[18]See note 7.

[19]The Leibniz notation expresses the idea of "summing." It is consistent with methods of approximation going back to antiquity. Later, building on the work of Cauchy and others, Riemann (1826–1866) worked out a definition of area between the graph of a function and the x-axis. In contemporary calculus textbooks, we usually find the Fundamental Theorem of Calculus presented following an introduction to Riemann integration. Unfortunately, that approach distracts students from being able to get the key insight that leads to the Fundamental Theorem of Calculus, a result that historically was known decades prior to Riemann's work. There are distinct questions. The Fundamental

area, each of which is of height $f(t)$ and width dt. The symbol "$\int$" is an elongated "S", for "Sum."

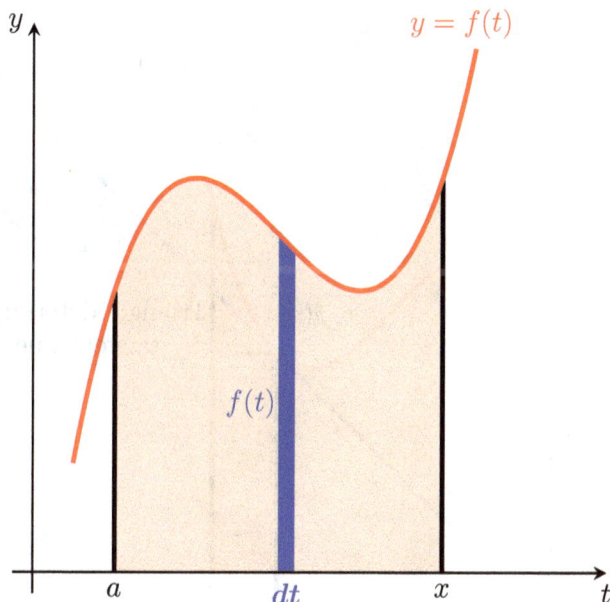

Figure 1.8   Leibnitz heuristics and notation: Area $A(x)$ under graph is a sum of areas of the form "$f(t)dt$": $A(x) = \int_a^x f(t)dt$.

The Question: For the area function $A(x) = \int_a^x f(t)dt$, what is the rate of change of the area, relative to changes in length x of the base? That is, what is

$$\frac{d}{dx}(A(x)) = \frac{d}{dx}\left(\int_a^x f(t)dt\right)?$$

Caution: In some books, $A(x) = \int_a^x f(t)dt$ is *defined* to be the anti-derivative of $y = f(x)$. This makes the issue a tautology. But the definition does not cause any problems. Why not? Because it is a consequence of the Fundamental Theorem of Calculus.

Theorem of Calculus answers the question: **What is the rate of change of area?** Riemann's work provided an answer to the question: **What is area?** Derivation of the Fundamental Theorem of Calculus from the definition of the Riemann integral answers the question: **Is Riemann's definition consistent with the earlier result about rate of change of area?**

Just as with the rolls of paper and the square, to get the key insight, we can again appeal to a diagram.

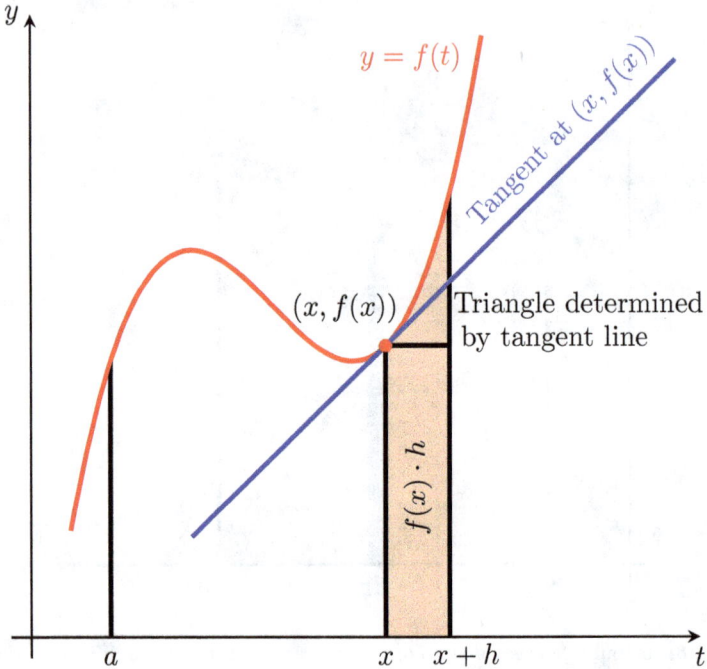

Figure 1.9 $A(x + h) - A(x) \approx f(x) \cdot h +$ [area of small triangle]. Small triangle is determined by the tangent line at $(x, f(x))$.

From the diagram,

$$\frac{\text{change in area}}{\text{change in length of base}}$$

$$= \frac{f(x) \cdot h + (\text{area of a small approximately triangular area})}{(h)}.$$

Appealing to the results of Section 1.1, area of the triangular piece is approximately

$$\frac{1}{2} \left(\text{base of the small triangle} \right) \times \left(\text{height of the small triangle} \right)$$

$$\approx \frac{1}{2} (h) \left(f'(x) \, h \right).$$

And so,

$$\frac{\text{change in area}}{\text{change in length of base}} = \frac{f(x) \cdot h + \left(\frac{1}{2}h\left(f'(x)h\right)\right)}{(h)}$$

$$= f(x) + \frac{1}{2}\left(f'(x)h\right).$$

Remember that x is fixed. As the change in length of the base approaches zero, the average rate of change of area approaches $f(x)$.

Geometrically, $f(x)$ is the length of the front-line of the advancing area.

This the key insight for the Fundamental Theorem:

Theorem 1.1 (Fundamental Theorem of Calculus).

Let $A(x) = \displaystyle\int_a^x f(t)dt$ represent the area between the graph of the function $y = f(x)$ and the x-axis. Then the rate of change of $A(x) = \displaystyle\int_a^x f(t)dt$ is the length of the advancing (or retreating) front-line $f(x)$.

Using Leibniz's notation, every symbol has a meaning when we write

$$\frac{d}{dx}\left(\int_a^x f(t)dt\right) = f(x).$$

Exercises

In Exercises 1.2.1–1.2.4, (a) find y as a function of x and (b) demonstrate the Fundamental Theorem of Calculus by differentiating the result of part (a).

Exercise 1.2.1

$$y = \int_0^x (3t + 8)\,dt.$$

Exercise 1.2.2

$$y = \int_{-1}^x 3t^2(t^3 + 1)^4\,dt.$$

Exercise 1.2.3

$$y = \int_1^x \left(t + \frac{1}{t}\right)dt.$$

Exercise 1.2.4

$$y = \int_{\pi/4}^x \sec^2 t\,dt.$$

In Exercise 1.2.5–1.2.6, use the Fundamental Theorem of Calculus to find $\dfrac{dy}{dx}$.

Exercise 1.2.5

$$y = \int_0^x e^{t^2} \, dt.$$

Exercise 1.2.6

$$y = \int_1^x \left(\frac{1}{t^4 + 1} \right) dt$$

In Exercise 1.2.7–1.2.8, express the solution of the initial value problem as an integral.

Exercise 1.2.7

$$\frac{dy}{dx} = \tan x, \; y(0) = 1.$$

Exercise 1.2.8

$$\frac{dy}{dx} = \sqrt{1 + x^3}, \; y(1) = 3.$$

1.3 Applications by which the theorem earns the name "fundamental"

To begin to see why the fundamental theorem of calculus is "fundamental," we look to the historical context.

In antiquity, only a few formulas for areas were known. Each area had its own definition. The area of a rectangle is

$$\Big(base\Big) \times \Big(height\Big).$$

By subtracting and adding congruent elements, the area of a parallelogram is (defined to be) the area of a rectangle of the same base and height; and the area of a triangle is one half that of a parallelogram of the same base and height. Hence, the area of a triangle is

$$\frac{1}{2} \left[\Big(base\Big) \times \Big(height\Big) \right].$$

And so on, that is, formulas can be obtained for areas of figures, when the figures can be constructed from rectangles, parallelograms and triangles. Partly because they all are reducible to rectangles, the answers are mutually compatible.

However, what is the area of a circle of radius r? This doesn't fit into the system so easily. We can make a beginning by observing that a planar circle is defined by a center C and its radius r. We can then use that radius to construct 4 squares to cover the circle. By insight into the diagram, the area of four squares each of length r is greater than the area of the circle of radius r.

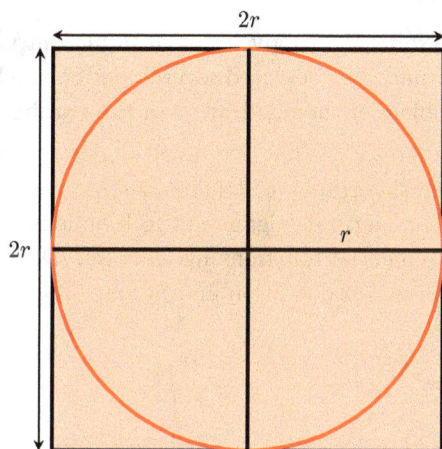

Figure 1.10   Area of circle of radius r is less than $4r^2$.

On the other hand, with more refined geometric constructions (using inscribed and circumscribed polygons[20]), the area of 3 squares of length r is less than the area of the circle. And so we get the following:

$$3r^2 < \left(\text{area of a circle of radius } r \right) < 4r^2.$$

Observe that the argument does not depend on the radius. And so we conclude that there is a ratio, call it π, $3 < \pi < 4$, such that the

$$\left(\text{area of a circle of radius } r \right) = \pi r^2.^{[21]}$$

There are many other figures in Euclidean geometry. Archimedes solved the problem of "quadrature of the parabola." He obtained a formula for the area of a parabola without needing to introduce unknown constants like 'π.' By subtraction, we then also get a formula for area between a parabola $y = x^2$ and the x-axis. To get his result, however, Archimedes

---

[20]See note 9.

[21]Using regular 96-gons, Archimedes obtained the sharp estimate $\dfrac{223}{7} < \pi < \dfrac{22}{7}$. Accurate estimates were also obtained in other cultures including, for instance, in astronomical calculations in the Shatapatha Brahmana, written in the 6th century B.C.E., in India.

had to develop the theory of non-terminating geometric series. He needed and reached an understanding of "limit of a sequence of geometric sums." [22] Note that all of this was done about 2000 years before equivalent results were (re-) discovered by European mathematicians. [23]

If there had been Nobel Prizes given out in in ancient Greece, we venture that Archimedes would have deserved several, one of which would have been for solving the problem of the quadrature of the parabola.

What about other areas? What are areas under the graphs of $y = x^3$, $y = x^4$, $\cdots$; under the graphs of trigonometric functions like $y = \sin x$, $y = \cos x$, $y = \tan x$; derived trigonometric functions like $y = \sec^2 x$ and $y = \csc^2 x$; and exponential functions like $y = 2^x$? Does each type of type of function require its own definition and perhaps its own Nobel Prize?

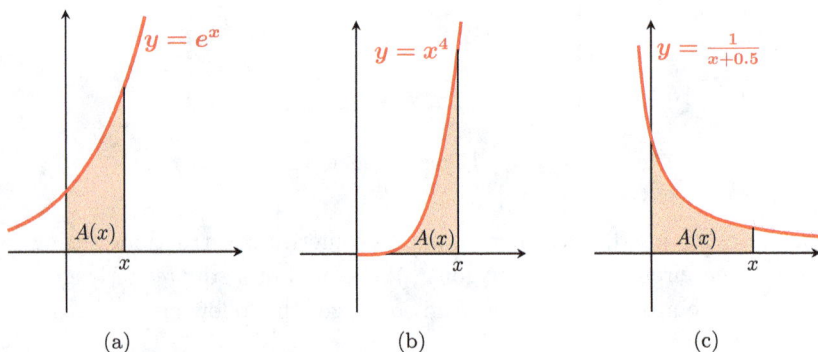

Figure 1.11   The Fundamental Theorem of Calculus provides the basis for computing areas. (a) $A(x) = e^x$. (b) $A(x) = x^5/5$. (c) $A(x) = ln(x + 0.5)$.

We begin to see the importance of the Fundamental Theorem of Calculus by observing that it allows us to reproduce Archimedes' solution, and solve many other such problems, almost effortlessly.

Let $A(x)$ be the area under the graph of the parabola given by $y = f(x) = x^2$, over the interval $[0, x]$. By the Fundamental Theorem of Calculus,

$$\frac{dA}{dx} = x^2.$$

[22]T. L. Heath, "Quadrature of the Parabola," in The Works of Archimedes (Cambridge: Cambridge University Press, 1897), Book I, 233–252, https://archive.org/details/worksofarchimede029517mbp/page/n421.

[23]Cauchy (1789–1857) was one of the European mathematicians who broke through to a definition. See Victor J. Katz, *A History of Mathematics, An Introduction*, 2nd ed. (Reading, MA: Addison Wesley Longman, Inc., 1998), 706–729.

Appealing to our repertoire of derivatives and anti-derivatives, we get that

$$A(x) = \frac{x^3}{3}.$$

Exactly similar arguments provide areas under the graphs of the other functions mentioned above, and many more. That is, in each case, the possibility of obtaining an explicit formula for area depends on having techniques for finding anti-derivatives.

Exercises

Exercise 1.3.1

Graph $y = f(t) = t^2$, $t \geq 0$.
The problem is to find the area under the graph of the parabola. As explained in the text, Archimedes solved the problem by developing the theory of convergent geometric series. Using derivatives, the solution to the problem is elegant and extends to an indefinite range of other functions and applications.

(a) Draw the area under the graph of $y = f(t) = t^2$, for the interval $t = 0$ to $t = 5$. What is the length of the vertical front-line?

(b) Without doing "calculus", what is an approximation for the area swept out, if the vertical front-line moves forward by $\Delta t = 0.1$?

(c) What is an approximation for the area swept out if the front-line moves forward by "small" $\Delta t = h > 0$?

(d) What is the instantaneous rate of change of area at $t = 5$?

(e) Repeat (c)-(d) for the interval $t = 0$ to $t = x$. That is, using first principles, what is the instantaneous rate of change of $A(x)$, the area under the graph of $y = f(t) = t^2$, at $t = x$? In other words, what is
$$\frac{d}{dx}A(x)?$$

(f) Using (e), what is the formula for the area $A(x)$ under the graph of $y = t^2$, $t \geq 0$, at $t = x$? This reproduces Archimedes' result.

Exercise 1.3.2

Repeat Exercise 1.3.1 for the areas under the graphs of the following functions:

(a) $y = t^4$, on the interval $t \geq 0$

(b) $y = \cos t$, on the interval $0 \le t \le \pi/2$

(c) $y = \sin t$, on the interval $0 \le t \le \pi/2$

(d) $y = \sec^2 t$, on the interval $0 \le t < \pi/2$

(e) $y = \dfrac{1}{t+1}$, on the interval $t \ge 0$

(f) $y = e^t$, on the interval $t \ge 0$

1.4   Linear approximation and area

Let $A(x)$ be the area under the graph of a function $y = f(x) > 0$, over the interval $[a, b]$.

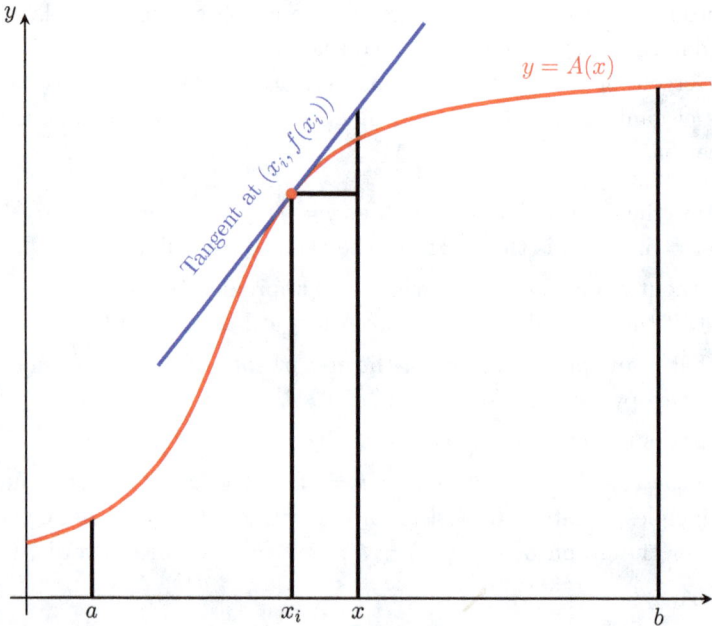

Figure 1.12   Linear approximation of area $A(x)$: $A(x) \approx A(x_i) + A'(x_i)(x - x_i)$.

An area function $A(x)$ is a function of x that can be graphed on an (x, y) plane. The function has its own derivative function, $A'(x)$. Assuming that the derivative exists everywhere, at each reference point x_i there is a linear approximation, namely,

$$A(x) \approx A(x_i) + A'(x_i) \cdot (x - x_i).$$

Equivalently, we can write

$$A(x) - A(x_i) \approx A'(x_i) \cdot (x - x_i).$$

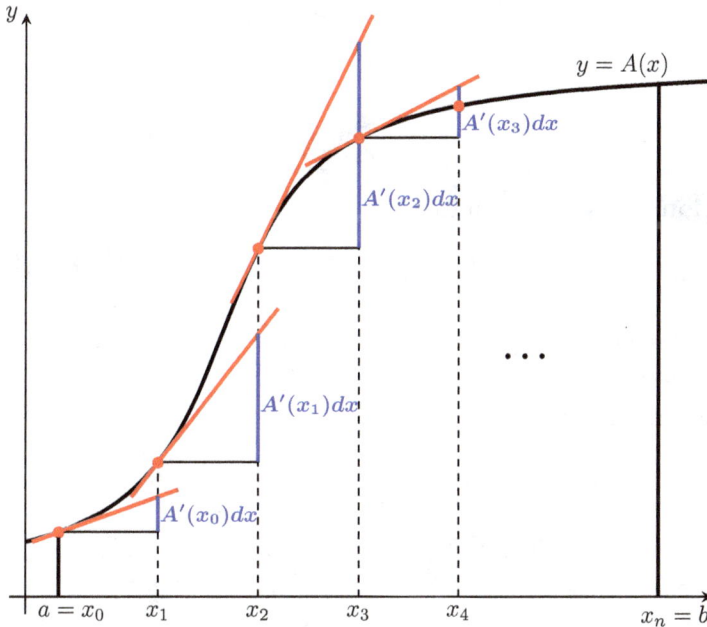

Figure 1.13 For each increment, $A(x_{i+1}) - A(x_i) \approx A'(x_i)dx$, where $dx = x_{i+1} - x_i$ for all $i = 1, 2, \cdots, n$. Therefore the net difference $A(b) - A(a)$ is approximated by the sum of approximations to incremental differences. That is, $A(b) - A(a) = \sum_{i=1}^{n-1} A'(x_i)dx$.

Partition $[a, b]$ into sub-intervals of equal length,

$$a = x_0 < x_1 < \cdots < x_{n-1} < x_n = b,$$

where $x_i - x_{i-1} = dx > 0$ for $i = 1, \cdots n$.

We get the following approximations:

$$A(x_n) - A(x_{n-1}) \approx A'(x_{n-1}) \cdot (x_n - x_{n-1}) = A'(x_{n-1})dx$$
$$A(x_{n-1}) - A(x_{n-2}) \approx A'(x_{n-2}) \cdot (x_{n-1} - x_{n-2}) = A'(x_{n-2})dx$$

$$\vdots \qquad\qquad \vdots$$

$$A(x_1) - A(x_0) \approx A'(x_0) \cdot (x_1 - x_0) = A'(x_0)dx$$

Adding equations, the left-hand-side telescopes. Recall also that $a = x_0$ and $x_n = b$.

So we get

$$A(b) - A(a) \approx \sum_{i=1}^{n-1} A'(x_i)dx. \tag{1.1}$$

If we let $dx \to 0$, and if the limit exists, it follows that

$$A(b) - A(a) = \lim_{dx \to 0} \sum_{i=1}^{n-1} A'(x_i)dx.$$

Question: Does this mean that

$$A(b) - A(a) = \lim_{dx \to 0} \sum_{i=1}^{n-1} A'(x_i)dx = \int_a^b A'(x)dx = \int_a^b f(x)dx?$$

So, does the Fundamental Theorem of Calculus really only depend on linear approximation?

Discussion: The first two equalities,

$$A(b) - A(a) = \lim_{dx \to 0} \sum_{i=1}^{n-1} A'(x_i)dx = \int_a^b A'(x)dx,$$

cause no serious difficulty. The second is consistent with the meaning of Leibniz's notation. And, the first equality follows from linear approximation, a refinement on techniques going back to ancient times. As we have just done, using linear approximation,

$$\sum_{i=1}^{n-1} A'(x_i)dx$$

approximates area under the graph of a function, a function that in the present context happens to be $y = A'(x)$.

In the Question, it is the third equality where we find what might seem to be "sleight of hand." Why can we write

$$\int_a^b A'(x)dx = \int_a^b f(x)dx?$$

It is here that we invoke the Fundamental Theorem of Calculus, namely, that $A'(x) = f(x)$.

With that identification, it is evident that the chain of equalities

$$A(b) - A(a) = \lim_{dx \to 0} \sum_{i=1}^{n-1} A'(x_i)dx = \int_a^b A'(x)dx = \int_a^b f(x)dx \tag{1.2}$$

is partly, but not merely, a consequence of linear approximation. That is, the ingredient that provides the third equality is the fact that the derivative of area under the graph of the function $y = f(x) > 0$ is the length of the front-line of advancing area.

Exercises

Exercise 1.4.1

Using equation 1.2, find the formula for the area under the graph of $y = f(x) = x$, over the interval $0 \leq x \leq b$. If you draw the graph, you will find that you already know the answer from classical geometry. The exercise here is to find it using the new method. *Hint*: Use subintervals of width $dx = \dfrac{1}{n}$. Also, what is $1 + 2 + \cdots + n$?

Exercise 1.4.2

Using equation 1.2, find the formula for the area under the graph of $y = f(x) = x^2$, over the interval $0 \leq x \leq b$. *Hint*: Use subintervals of width $dx = \dfrac{1}{n}$. Also, what is $1^2 + 2^2 + \cdots + n^2$? One way to find this sum is to construct the following series of equations:
$(1+1)^3 - 1^3 = 3 \cdot 1^2 + 3 \cdot 1 + 1$
$(2+1)^3 - 2^3 = 3 \cdot 2^2 + 3 \cdot 2 + 1$
$\cdots$
$((n-1)+1)^3 - (n-1)^3 = 3 \cdot (n-1)^2 + 3 \cdot (n-1) + 1.$
Adding, the left side is a telescoping series while the right side can be written in terms of the unknown $1^2 + 2^2 + \cdots + n^2$ along with two sums that are already known.

Exercise 1.4.3

Generalize Exercises 1.4.1 and 1.4.2. That is, for k a positive integer, using equation 1.2 of Section 1.4, find the formula for the area under the graph of $y = f(x) = x^k$, over the interval $0 \leq x \leq b$. *Hint*: What is $1^k + 2^k + \cdots + n^k$? Imitate the computations used in Exercises 1.4.1 and 1.4.2. You will need the binomial theorem.

1.5   Intimations of Green's theorem in one dimension

Equations 1.1 and 1.2 foreshadow computations needed in developing Green's theorem, Stokes' theorem, and the divergence theorem.

For the vector theorems treated in this book, contexts differ. However, in all cases, a crucial computation involves essentially the same strategy. What follows is a pared down and simplified illustration:

Suppose that $g(x, y)$ is defined and differentiable on the rectangular region

$$R = [x, x + dx] \times [y, y + dy]$$

and that, for each x, $g(x, y)$ is a (linear mass density)[24] function in y. This means that $g(x, y)dy$ approximates the mass of a length dy in the y direction, based at (x, y).

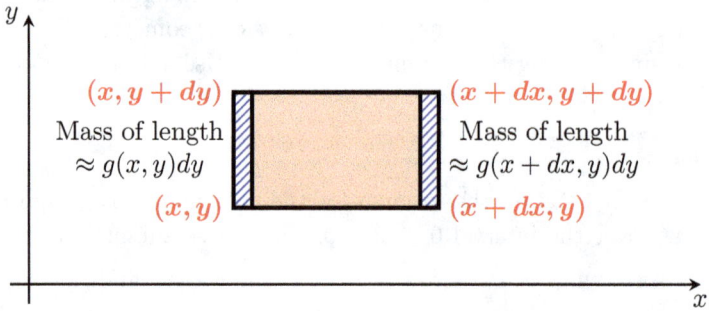

Figure 1.14   Shaded verticals are to indicate that there is a linear mass-function in the y-direction.

Now consider the difference

$$g(x + dx, y)dy - g(x, y)dy,$$

with g defined as above.

By definition of $g(x, y)$, this approximates the difference in mass of the two vertical sides of the rectangular region R.

By linear approximation, we can write the following:

$$
\begin{aligned}
g(x + dx, y)dy - g(x, y)dy &= [g(x + dx, y) - g(x, y)]\, dy \\
&= [g_x(x, y)dx]\, dy \\
&= g_x(x, y)(dx dy) \\
&= g_x(x, y)(\text{area of rectangle } R) \\
&= [\text{weighted area element}]
\end{aligned}
$$

---

[24]These will be introduced in the main text, in Parts 1 and 2.

In other words,

$$g(x + dx, y)dy - g(x, y)dy = g_x(x, y) \, (\text{area of rectangle } R) \, .$$

And so: A difference of two weighted line elements in the y direction (each of which is an integrand for a one-dimensional integral) is approximated by a weighted area element, an integrand for a two-dimensional integral. Moreover, the weight for the area element is a derivative of the weight for the line element.

This computation does not provide the fundamental theorems of vector calculus. But you might keep it in mind as foreshadowing for the development of the vector calculus theorems.

Exercises

Exercise 1.5.1
Let $g(x, y) = x^2 y^3$. Using linear approximation, show

$$g(x + dx, y)dy - g(x, y)dy \approx 2xy^3(dxdy) = g_x(x, y)(\text{area of rectangle})$$

In Exercises 1.5.2–1.5.4, compute the first-order partial derivatives.

Exercise 1.5.2
$f(x, y) = 2x^3 + 3x^2 y^3 - y^4 + 8$

Exercise 1.5.3
$f(x, y) = ye^{xy} + 10xy^2$

Exercise 1.5.4
$h(x, y, z) = \dfrac{xyz}{x^2 + y^2 + z^2}$

Exercise 1.5.5
For $z = x^2 + 3ye^{x/y}$, find an equation of the tangent plane at $P = (0, 1, 3)$.

Exercise 1.5.6
The linearization of a function $z = f(x, y)$ at a point (x_0, y_0) is the function $L(x, y)$: (a) that is linear in both x and y; (b) that agrees with $f(x, y)$ at the reference point (x_0, y_0); and (c) whose partial derivatives also agree at the reference point. In other words, the linearization is of the form

$$z = L(x, y) = ax + by + f(x_0, y_0)$$

and satsifies

$$L(x_0, y_0) = f(x_0, \ y_0), \ a = \frac{\partial}{\partial x} f(x_0, y_0), \ b = \frac{\partial}{\partial y} f(x_0, y_0).$$

Find the linearization $L(x, y)$ of

$$f(x, y) = \frac{x}{1 + y^2}$$

at $(2, 1)$ and use it to estimate $f(2.1, 1.1)$.

Exercise 1.5.7

Suppose that a function $z = f(x, y)$ is defined on a neighborhood containing the reference point (x_0, y_0). Recall that the *differential* dz is the linear approximation to $f(x, y) - f(x_0, y_0)$, in terms of $dx = (x - x_0)$ and $dy = (y - y_0)$. Find the differential dz for $z = xy + e^{xy}$.

1.6   Dot-product or scalar product: a solution to a problem about projection length, in coordinates

You probably already know the formula for the "dot product," learned perhaps in Calculus III, linear algebra, or some other undergraduate course. But what is the *dot product*? Much of geometry in undergraduate mathematics is Euclidean geometry and yet the dot product does not appear in Euclid's geometry.

At a time when coordinate charts were in common use for making maps for surveying and traveling, similar advances in mathematics were made by both Fermat and Descartes. In Euclid's geometry there are abstractly defined figures, points, lengths and so on. In coordinate geometry, one provides figures with locations and dimensions on a coordinate plane. It is still Euclidean geometry. But as history shows, the new approach is enormously fruitful. Among other things, it allows one to solve geometry problems by doing algebra and to solve algebra problems by appealing to geometry. It is in this context that the following question arises:

Question: Given two vectors $\mathbf{v} = (a, b)$ and $\mathbf{w} = (c, d)$ emanating from the same point, in terms of the coordinates a, b, c, and d, what is the length of the orthogonal projection of $\mathbf{v}$ onto the line that determined by vector $\mathbf{w}$?

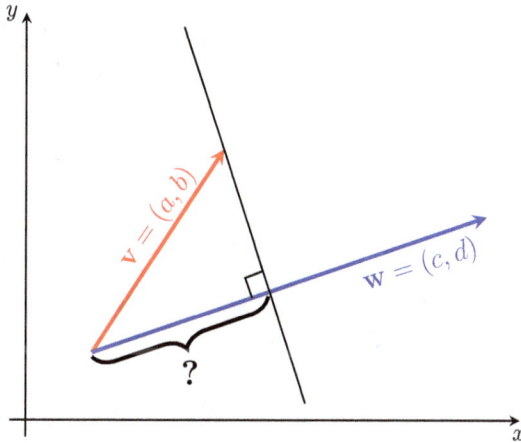

Figure 1.15   Determine the length of the orthogonal projection of **v** onto **w** in terms of coordinates a, b, c and d.

Problem 1.1.

(i) Assuming the Pythagorean Theorem and appealing only to the axioms of Euclidean geometry, develop the law of cosines. *Hint*: Extend the diagram so that you can express the problem in terms of two nested right angle triangles.

(ii) Apply the law of cosines to the case where two sides of a triangle are given by coordinate vectors $\mathbf{v} = (a, b)$ and $\mathbf{w} = (c, d)$ emanating from the origin. Depending on one's choice of initial and final point, a third side of the triangle is $\mathbf{u} = \pm(c - a, d - b)$.

(iii) Develop a formula for the magnitude of the projection of $\mathbf{v} = (a, b)$ onto $\mathbf{w} = (c, d)$, in terms of the coordinates of $\mathbf{v} = (a, b)$ and $\mathbf{w} = (c, d)$. *Hint*: See ii.

(iv) Observe that a key element of the projection formula obtained in iii. is the combination $ac + bd$. This combination is called the *dot product* and is written $\mathbf{v} \cdot \mathbf{w} = (a, b) \cdot (c, d) = ac + bd$.

(v) Recast your results so that you get the well-known formula for $\mathbf{v} \cdot \mathbf{w}$ in terms of $\| \mathbf{v} \|$, $\| \mathbf{w} \|$ and $\cos \theta$, where θ is an angle between $\mathbf{v} = (a, b)$ and $\mathbf{w} = (c, d)$.

Exercises

In Exercises 1.6.1–1.6.3, compute $\mathbf{u} \cdot \mathbf{v}$, $\| \mathbf{u} \|$, and $\| \mathbf{v} \|$.

Exercise 1.6.1

$\mathbf{u} = (3, 1)$, $\mathbf{v} = (4, -2)$

Exercise 1.6.2

$\mathbf{u} = (-1, 3, 1)$, $\mathbf{v} = (0, 7, 2)$

Exercise 1.6.3

$\mathbf{u} = 6\mathbf{i} - \mathbf{j} + 2\mathbf{k}$, $\mathbf{v} = 4\mathbf{i} + \mathbf{j} - 3\mathbf{k}$

Exercise 1.6.4

Find a unit vector that points in the same direction as the vector $\mathbf{u} = \mathbf{i} - 2\mathbf{j} + 3\mathbf{k}$.

In Exercises 1.6.5–1.6.6, find the angle between $\mathbf{u}$ and $\mathbf{v}$.

Exercise 1.6.5

$\mathbf{u} = \mathbf{i} + 3\mathbf{j}$, $\mathbf{v} = -2\mathbf{i} + 3\mathbf{j}$

Exercise 1.6.6

$\mathbf{u} = \mathbf{i} + \mathbf{j} + \mathbf{k}$, $\mathbf{v} = 2\mathbf{i} + 5\mathbf{j} - \mathbf{k}$

Exercise 1.6.7

Let $\mathbf{v} = (a, b)$ and $\mathbf{e} = (1, 0)$. From classical geometry, $\mathbf{v} \cdot (1, 0) = a = \|\mathbf{v}\| \cos \theta$ is the projection of $\mathbf{v} = (a, b)$ onto the x-axis. Show that if $\mathbf{u} = (c, d)$ is any unit vector, then $\mathbf{v} \cdot \mathbf{u}$ is the projection of $\mathbf{v}$ onto the u-axis.

In Exercises 1.6.8–1.6.9, calculate the projection of $\mathbf{u}$ onto $\mathbf{v}$.

Exercise 1.6.8

$\mathbf{u} = \mathbf{i} + \sqrt{3}\mathbf{j}$, $\mathbf{v} = -2\mathbf{i} - \sqrt{3}\mathbf{j}$

Exercise 1.6.9

$\mathbf{u} = 2\mathbf{i} - \mathbf{j} + 3\mathbf{k}$, $\mathbf{v} = -\mathbf{i} + \mathbf{j} + 3\mathbf{k}$

1.7   Areas and volumes, in coordinates

The problem of rewriting classical geometry in terms of coordinates continues.

Areas

Two vectors $\mathbf{v} = (a_1, a_2)$ and $\mathbf{w} = (b_1, b_2)$ determine sides of a parallelogram.

Question: What is the area of the parallelogram, in terms of the coordinates of $\mathbf{v}$ and $\mathbf{w}$?

> **Problem 1.2.**
> The classical formula for the area of a parallelogram is (base) $\times$ (height). To compute height, we need the sine function. Starting with the classical formula and using both the sine function and the Pythagorean formula, show from first principles that the area of the parallelogram determined by $\mathbf{v} = (a_1, a_2)$ and $\mathbf{w} = (b_1, b_2)$ is $|a_1 b_2 - b_1 a_2|$. *Hint:* Express the sine function in terms of the cosine function and use Problem 1.1.

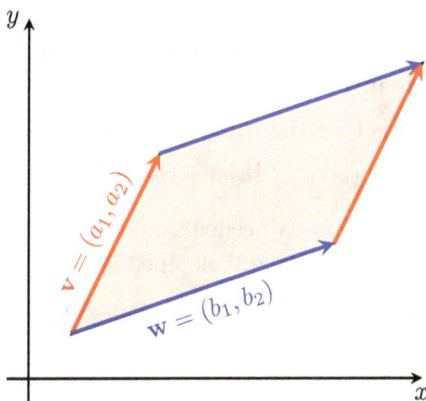

Figure 1.16   Determine the area of the parallelogram determined by $\mathbf{v} = (a_1, a_2)$ and $\mathbf{w} = (b_1, b_2)$, in terms of coordinates a_1, a_2, b_1 and b_2.

The quantity $a_1 b_2 - b_1 a_2$ is called the determinant of the pair of vectors $\mathbf{v}$ and $\mathbf{w}$. The determinant can be written as follows:

$$det \begin{pmatrix} a_1 & a_2 \\ b_1 & b_2 \end{pmatrix} = a_1 b_2 - b_1 a_2.$$

Also common notation is using vertical lines:

$$det \begin{pmatrix} a_1 & a_2 \\ b_1 & b_2 \end{pmatrix} = \begin{vmatrix} a_1 & a_2 \\ b_1 & b_2 \end{vmatrix} = a_1 b_2 - b_1 a_2.$$

Volumes

In (x, y, z) space, three vectors $\mathbf{u} = (a_1, a_2, a_3)$, $\mathbf{v} = (b_1, b_2, b_3)$ and $\mathbf{w} = (c_1, c_2, c_3)$ determine sides of a parallelepiped.

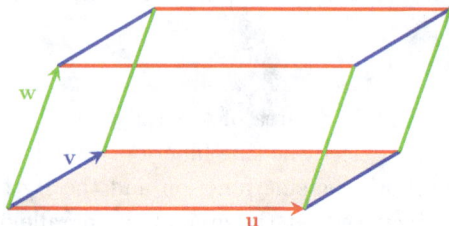

Figure 1.17   Determine the volume of a parallelepiped determined by $\mathbf{u} = (a_1, a_2, a_3)$, $\mathbf{v} = (b_1, b_2, b_3)$ and $\mathbf{w} = (c_1, c_2, c_3)$, in terms of the coordinates a_1, a_2, a_3, b_1, b_2, b_3, c_1, c_2, and c_3.

Question: What is the volume of the parallelepiped, in terms of the coordinates of the vectors $\mathbf{u}$, $\mathbf{v}$ and $\mathbf{w}$?

The classical formula is that the volume of a parallelepiped is

$$(\text{area of base}) \times (\text{height}).$$

Invoke the previous problem as needed and so work out that, in terms of coordinates, the volume of the parallelepiped is

$$\left| [a_1 b_2 c_3 + a_2 b_3 c_1 + a_3 b_1 c_2] - [c_1 b_2 a_3 + c_2 b_3 a_1 + c_3 b_1 a_2] \right|.$$

The quantity

$$[a_1 b_2 c_3 + a_2 b_3 c_1 + a_3 b_1 c_2] - [c_1 b_2 a_3 + c_2 b_3 a_1 + c_3 b_1 a_2]$$

is called the *determinant* of the three vectors $\mathbf{u} = (a_1, a_2, a_3)$, $\mathbf{v} = (b_1, b_2, b_3)$ and $\mathbf{w} = (c_1, c_2, c_3)$.

Similar to the 2-d case, this is often written:

$$det \begin{pmatrix} a_1 & a_2 & a_3 \\ b_1 & b_2 & b_3 \\ c_1 & c_2 & c_3 \end{pmatrix} = \begin{vmatrix} a_1 & a_2 & a_3 \\ b_1 & b_2 & b_3 \\ c_1 & c_2 & c_3 \end{vmatrix}$$

$$= [a_1 b_2 c_3 + a_2 b_3 c_1 + a_3 b_1 c_2] - [c_1 b_2 a_3 + c_2 b_3 a_1 + c_3 b_1 a_2].$$

For students familiar with linear algebra

Evidently, the two cases of "determinant" are similar. We can begin to explore that similarity by looking to underlying geometry.

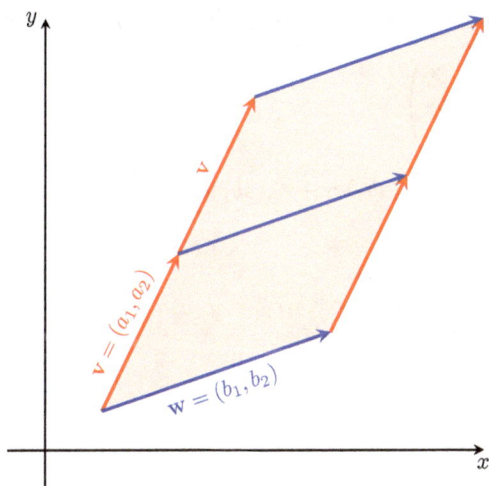

Figure 1.18 Area of parallelogram determined by $2\mathbf{v}$ and $\mathbf{w}$ is twice the area of that determined by $\mathbf{v}$ and $\mathbf{w}$.

If one edge of a parallelogram is multiplied by a factor of 2, say, then the area is also multiplied by a factor of 2.

In the determinant formula,

$$det \begin{pmatrix} a_1 & a_2 \\ 2b_1 & 2b_2 \end{pmatrix} = a_1(2b_2) - (2b_1)a_2$$

$$= 2(a_1 b_2 - b_1 a_2)$$

$$= 2\, det \begin{pmatrix} a_1 & a_2 \\ b_1 & b_2 \end{pmatrix}.$$

The same sort of result holds for volume of a parallelepiped. For instance, from classical geometry, the volume determined by three vectors

$$\mathbf{u} = (a_1, a_2, a_3),$$
$$4\mathbf{v} = (4b_1, 4b_2, 4b_3) \text{ and}$$
$$\mathbf{w} = (c_1, c_2, c_3)$$

is four times the volume determined by $\mathbf{u} = (a_1, a_2, a_3)$, $\mathbf{v} = (b_1, b_2, b_3)$ and $\mathbf{w} = (c_1, c_2, c_3)$. And, if we substitute the coordinates of $\mathbf{u} = (a_1, a_2, a_3)$,

$4\mathbf{v} = (4b_1, 4b_2, 4b_3)$ and $\mathbf{w} = (c_1, c_2, c_3)$ into the 3-d determinant we get

$$det \begin{pmatrix} a_1 & a_2 & a_3 \\ 4b_1 & 4b_2 & 4b_3 \\ c_1 & c_2 & c_3 \end{pmatrix}$$

$$= \Big[a_1(4b_2)c_3 + a_2(4b_3)c_1 + a_3(4b_1)c_2 \Big] - \Big[c_1(4b_2)a_3 + c_2(4b_3)a_1$$
$$+ c_3(4b_1)a_2 \Big]$$

$$= 4\Big\{ [a_1b_2c_3 + a_2b_3c_1 + a_3b_1c_2] - [c_1b_2a_3 + c_2b_3a_1 + c_3b_1a_2] \Big\}$$

$$= 4\,det \begin{pmatrix} a_1 & a_2 & a_3 \\ b_1 & b_2 & b_3 \\ c_1 & c_2 & c_3 \end{pmatrix}$$

Problem 1.3.

Make a diagram for the volumes discussed above.

If we switch the order of two sides then areas and volumes do not change. However, the two vectors representing the sides change in relative orientation.

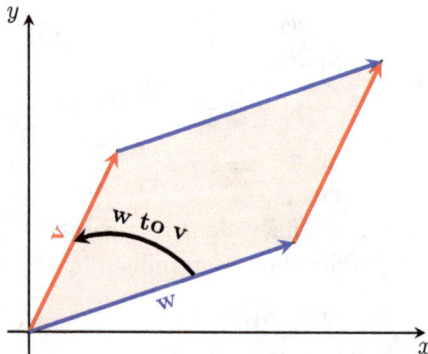

Figure 1.19   In the diagram, $\mathbf{v}$ to $\mathbf{w}$ is a negative rotation, while $\mathbf{w}$ to $\mathbf{v}$ is a positive rotation.

This shows up in the determinant formula as a change in sign, but leaves the absolute value invariant. In the 2-d case, the computation is as follows:

$$det \begin{pmatrix} a_2 & a_1 \\ b_2 & b_1 \end{pmatrix} = (a_2 b_1 - b_2 a_1) = -(a_1 b_2 - b_2 a_1) = -det \begin{pmatrix} a_1 & a_2 \\ b_1 & b_2 \end{pmatrix}.$$

Two properties common to both the 2-d and 3-d determinant functions are: they are linear in each vector component; and they alternate in sign when two vectors are switched.

Differential forms generalize areas and volumes to higher dimensions. For readers who plan to go on study differential forms, note that these two properties are all that is needed to define a *determinant* function in n dimensions. In other words, a *generalized n-dimensional determinant function (or n-dimensional volume function)* is any real-valued function

$$D : (\mathbf{v}_1, \mathbf{v}_2, \cdots, \mathbf{v}_n) \longrightarrow \mathbb{R}$$

with the following two properties:

(i) D is linear in each component (multi-linear); and

(ii) when two vectors are switched, the value changes in sign (alternating). That is,
$$D(\mathbf{v}_1, \cdots, \mathbf{v}_i, \cdots, \mathbf{v}_j, \cdots, \mathbf{v}_n) = (-1)D(\mathbf{v}_1, \cdots, \mathbf{v}_j, \cdots, \mathbf{v}_i, \cdots, \mathbf{v}_n).$$

For any n, there are many such volume functions. As the following example illustrates, it is a matter of choosing units.

Problem 1.4.

With $\mathbf{e}_1 = (1, 0)$ and $\mathbf{e}_2 = (0, 1)$, define $D \begin{pmatrix} 1 & 0 \\ 0 & 1 \end{pmatrix} = 3$.

(i) Show that the defining properties determine $D \begin{pmatrix} a_1 & a_2 \\ b_1 & b_2 \end{pmatrix}$ for all pairs of vectors $\mathbf{v} = (a_1, a_2)$ and $\mathbf{w} = (b_1, b_2)$.

(ii) What is the relation between this $D \begin{pmatrix} a_1 & a_2 \\ b_1 & b_2 \end{pmatrix}$ and the classical
$$det \begin{pmatrix} a_1 & a_2 \\ b_1 & b_2 \end{pmatrix} = a_1 b_2 - b_1 a_2?$$

Problem 1.5.

Show that a generalized determinant function is completely determined by its value on "the unit n-cube," that is, by the value of $D(\mathbf{e}_1, \mathbf{e}_2, \cdots, \mathbf{e}_n)$, where $\mathbf{e}_1 = (1, 0, 0, \cdots, 0)$, $\mathbf{e}_2 = (0, 1, 0, \cdots, 0)$, $\cdots$, $\mathbf{e}_n = (0, 0, 0, \cdots, 1)$.

Exercises

In Exercises 1.7.1–1.7.2, compute $\mathbf{u} + \mathbf{v}$ and $2\mathbf{u} - 3\mathbf{v}$.

Exercise 1.7.1

$\mathbf{u} = (-1, 5)$, $\mathbf{v} = (2, -3)$

Exercise 1.7.2

$\mathbf{u} = (1, -1, 3)$, $\mathbf{v} = (7, 0, -1)$

Exercise 1.7.3

Find the area of the parallelogram determined by the vectors $\mathbf{u} = (-1, 5)$ and $\mathbf{v} = (2, -3)$.

Exercise 1.7.4

Find the volume of the parallelepiped determined by the vectors $\mathbf{u} = (2, -1, 3)$, $\mathbf{v} = (1, 3, 0)$, and $\mathbf{w} = (0, 6, 1)$.

1.8   Vector cross-product: a mathematical solution to a problem about rotation, from classical physics

Setting up the problem

A topic in undergraduate physics is rotational motion.

Imagine a rigid axle, and a rigid radial arm attached and orthogonal to the axle. The arm can only rotate in the plane Π that is orthogonal to the axle. The radial arm has a length and direction in (x, y, z) space and so can be represented by a "vector", namely, $\mathbf{r} = (a_1, a_2, a_3)$. Suppose $\mathbf{F} = (b_1, b_2, b_3)$ represents a force "applied" at the end-point of the radial arm. We can assume that the force vector is also in the plane Π. (If $\mathbf{F}$ is not in Π then, since the structure is assumed to be rigid, and in classical physics it is found experimentally that components of acceleration add linearly, we could use the projection of $\mathbf{F}$ onto the plane Π. Here, we don't need that generality.)

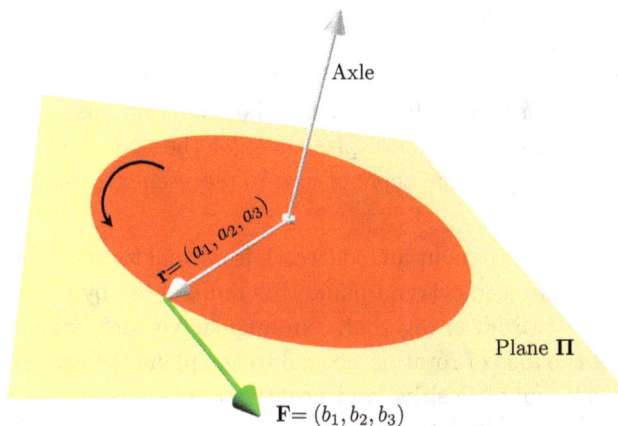

Figure 1.20   Rigid radial arm attached and orthogonal to an axle. Applied force $\mathbf{F} = (b_1, b_2, b_3)$ at end-point of $\mathbf{r} = (a_1, a_2, a_3)$ and coplanar with plane of rotation of $\mathbf{r}$ about axle.

The problem is to determine the angular acceleration of the radial arm caused by the applied force $\mathbf{F} = (b_1, b_2, b_3)$.

The structure is rigid. Within the constraints, only rotational motion is possible. By taking the component of $\mathbf{F} = (b_1, b_2, b_3)$ in the plane Π that is orthogonal to $\mathbf{r} = (a_1, a_2, a_3)$, and using the law of the lever, we can distinguish three cases:

(1) Angular acceleration is maximum when $\mathbf{F} = (b_1, b_2, b_3)$ is orthogonal to $\mathbf{r} = (a_1, a_2, a_3)$.

(2) Angular acceleration is zero when $\mathbf{F} = (b_1, b_2, b_3)$ is parallel to $\mathbf{r} = (a_1, a_2, a_3)$ (pointing either toward or away from the axis of rotation).

(3) Angular acceleration is between (i) and (ii), when $\mathbf{F} = (b_1, b_2, b_3)$ is at other angles relative to $\mathbf{r} = (a_1, a_2, a_3)$. More precisely, using the law of the lever, angular acceleration is proportional to the area of the parallelogram determined by the vectors $\mathbf{F} = (b_1, b_2, b_3)$ and $\mathbf{r} = (a_1, a_2, a_3)$.

Problem 1.6.
Make diagrams for cases (1), (2) and (3).

Angular vectors

In experiments, it is found that if two forces $\mathbf{F}_1$ and $\mathbf{F}_2$ are applied, angular accelerations add. However, angular accelerations are not vectors whose components represent displacement of the end-point of the radial arm. There is the question, then, of how to represent angular accelerations by quantities that add.

In physics, rotation is about an axle which is normal to the plane of motion. And, in coordinate geometry, a plane Π is represented by a normal vector. Bringing these two ideas together, "an angular vector" is defined to be a vector along the axis of rotation normal to the plane Π in which rotational displacement occurs. Positive and negative angular rotations in Π about the axle are represented by vectors pointing "upward" and "downward" along the axis (or axle) of rotation, respectively.[25]

This is well-known in classical physics and leads to the following mathematical problem:

To more easily allow for other applications, let's first replace the symbols $\mathbf{F}$ and $\mathbf{r}$ by $\mathbf{u}$ and $\mathbf{v}$.

Given $\mathbf{u} = (a_1, a_2, a_3)$, $\mathbf{v} = (b_1, b_2, b_3)$, find a vector $\mathbf{c} = (c_1, c_2, c_3)$ that has the following three properties:

(1) $\mathbf{c} = (c_1, c_2, c_3)$ is orthogonal to $\mathbf{u} = (a_1, a_2, a_3)$;

(2) $\mathbf{c} = (c_1, c_2, c_3)$ is orthogonal to $\mathbf{v} = (b_1, b_2, b_3)$; and

(3) $\| \mathbf{c} \| = \| (c_1, c_2, c_3) \| = $ (area of parallelogram determined by $\mathbf{u}$ and $\mathbf{v}$).

The new vector is called the *cross-product or vector product of* $\mathbf{u}$ and $\mathbf{v}$ and is denoted by the $\mathbf{u} \times \mathbf{v}$.

> **Problem 1.7.**
> Solve the system of equations (1), (2) and (3) and obtain that
>
> $$\mathbf{c} = (c_1, c_2, c_3) = \mathbf{u} \times \mathbf{v}$$
> $$= (a_1, a_2, a_3) \times (b_1, b_2, b_3)$$
> $$= (a_2 b_3 - a_3 b_2, a_3 b_1 - a_1 b_3, a_1 b_2 - a_2 b_1).$$

---

[25] For now, this is partly descriptive. But appealing to diagrams we get the right formulas.

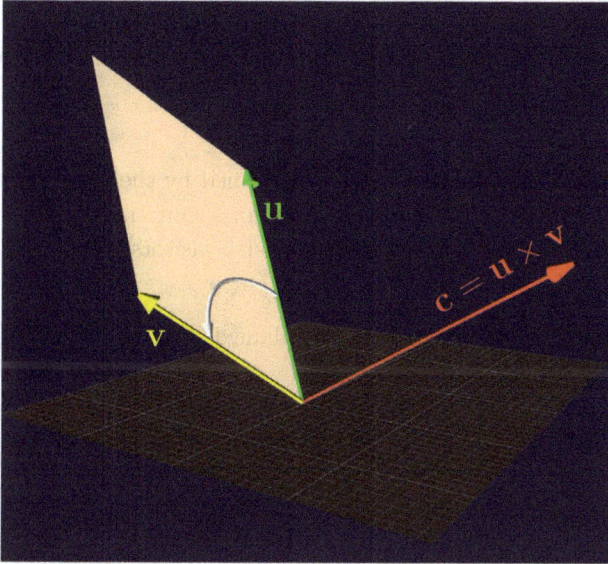

Figure 1.21 Given $\mathbf{u} = (a_1, a_2, a_3)$ and $\mathbf{v} = (b_1, b_2, b_3)$, the cross-product $\mathbf{c}$ is defined to be the vector that is orthogonal to both $\mathbf{u}$ and $\mathbf{v}$, whose length is equal to the area of the parallelogram determined by $\mathbf{u}$ and $\mathbf{v}$, and whose direction is determined by the "right-hand rule".

Exercises

In Exercises 1.8.1–1.8.2, compute the cross product of $\mathbf{u}$ and $\mathbf{v}$.

Exercise 1.8.1

$\mathbf{u} = (-3, 1, 2)$, $\mathbf{v} = (2, 0, -1)$

Exercise 1.8.2

$\mathbf{u} = 2\mathbf{i} - \mathbf{j} + \mathbf{k}$, $\mathbf{v} = \mathbf{i} + 4\mathbf{k}$

Exercise 1.8.3

Find a unit vector that is orthogonal to both $\mathbf{j} + \mathbf{k}$ and $-\mathbf{i} + \mathbf{j} - \mathbf{k}$.

Exercise 1.8.4

Assume that $\mathbf{a} \times \mathbf{b} = (0, 2, 1)$, $\mathbf{a} \times \mathbf{c} = (-1, 1, 2)$ and $\mathbf{b} \times \mathbf{c} = (1, 3, -2)$. Calculate $\mathbf{a} \times (2\mathbf{b} + \mathbf{c})$ and $(\mathbf{a} + \mathbf{b}) \times (3\mathbf{a} + \mathbf{b} - \mathbf{c})$.

Exercise 1.8.5

Find the area of the parallelogram determined by the three points $(1, 2, 3)$, $(1, 2, 4)$ and $(1, 1, 1)$.

Exercise 1.8.6

Find the area of the parallelogram determined by the three points $(1, 2, 3)$, $(2, 4, 6)$ and $(3, 6, 9)$. What does this mean about the relative location of these points? Can you prove the result using vectors?

Exercise 1.8.7

Find a vector that is orthogonal to the plane determined by the three points $(1, 0, 0)$, $(0, 1, 0)$ and $(0, 0, 1)$.

1.9   The chain rule in one variable

The chain rule is used throughout calculus, in all dimensions. It is a feature of linear approximation.

One dimension, without calculus

Example

The rule can be discovered in elementary measurement problems.

Consider the two ratios:

$$\frac{32 \text{ inches}}{1 \text{ yard}}$$

and

$$\frac{\frac{1}{3} \text{ yard}}{1 \text{ foot}}.$$

The question is whether or not it possible to use these given ratios to determine the ratio

$$\frac{\text{inches}}{\text{feet}}.$$

Of course, the ratio inches to feet is familiar and known. But the question here (a special case of the chain rule) is to use the given ratios — inches to yards, and yards to feet — to determine the ratio inches to feet.

Why is the question plausible?

The first ratio relates inches and yards. The second ratio relates yards and feet. There is, then, a link between the two ratios, a common unit, yards.

The trick is to take advantage of that common unit.

One way to do that is as follows:

$$\frac{32 \text{ inches}}{1 \text{ yard}} \times \frac{\frac{1}{3} \text{ yard}}{1 \text{ foot}} = \frac{12''}{\text{foot}}.$$

The common unit "yard" cancels and the arithmetic provides a ratio of "inches to feet," that is, $\frac{12''}{\text{foot}}$, or in the symbolism of proportions, a $12 : 1$ ratio.

Example

Suppose that a car gets highway mileage of

$$\frac{30 \text{ miles}}{\text{gallon}};$$

and that on a particular trip the car is consuming gasoline at a rate of

$$\frac{2 \text{ gallons}}{\text{hour}}.$$

How fast is the car traveling?

Again, an ad hoc solution is not difficult. But does not the approach of the previous example suggest the possibility of more a systematic solution?

We are looking for the ratio $\frac{\text{miles}}{\text{hour}}$.

We are given two ratios with units $\frac{\text{miles}}{\text{gallon}}$ and $\frac{\text{gallons}}{\text{hour}}$.

Clearly, the product of these two ratios will be in the required units, $\frac{\text{miles}}{\text{hour}}$.

In other words, the ratio is obtained as follows:

$$\frac{30 \text{ miles}}{\text{gallon}} \times \frac{2 \text{ gallons}}{\text{hour}} = \frac{60 \text{ miles}}{\text{hour}}.$$

More generally, ratios change. But that is precisely what calculus is designed to handle.

Example

Let $y = x^2$, $x = x(t) = 5t + 4$ and consider the composition

$$y(t) := y\big(x(t)\big) = (5t + 4)^2.$$

We have

$$\frac{dy}{dx} = 2x \quad \text{and} \quad \frac{dx}{dt} = 5.$$

The problem is to determine $\dfrac{dy}{dt}$, by using the two given ratios.

As in the previous example, one could solve this in an ad hoc way. For instance, in the present context, we could expand

$$y(t) := y\Big(x(t)\Big) = (5t + 4)^2 = 25t^2 + 40t + 16$$

and then apply derivative formulas to the resulting quadratic function. The question here, however, is different.

While we are only using elementary functions, the problem anticipates the case where functions involved are more sophisticated. They could be high degree polynomials, trigonometric functions, and so on.

And so the question is: Can we use the rates of change of the component functions to determine the rate of change of a composition function?

Since the composition function is built out of known component functions, does it not stand to reason that this might be possible? Part of the value of Leibniz's symbolism is that it reminds us that while the derivative of a function is not a ratio, it is a limit of ratios. And, indeed, using linear approximation, that limit can be used to estimate change in function values. And so, taking advantage of the Leibniz notation, might we not anticipate that

$$\left(\frac{dy}{dt}\right) = \left(\frac{dy}{dx}\right)\left(\frac{dx}{dt}\right)?$$

But again, the derivative symbols are limits of ratios. So, let's dig into these limits in order to get a better grip on this.

To get at $\left(\dfrac{dy}{dt}\right)$, we need to *estimate* how a small increment Δt impacts the composition

$$y(t) = f\left(x(t)\right) = (5t + 4)^2.$$

A value for t determines x; and a value for x determines y.

(1) For a small increment Δt, there is a small increment in $x = x(t) = 5t + 4$:

$$\Delta x = \left(\frac{dx}{dt}\right)\Delta t = 5\Delta t.$$

(2) For a small increment Δx, there is a small increment in $y = f(x) = x^2$:

$$\Delta y \approx \left(\frac{df}{dx}\right)\Delta x.$$

As we did in the last two examples, we can bring these ratios together. Algebraically, that means substituting (1) into (2). Hence,

$$\left(\frac{\Delta y}{\Delta t}\right) = \left(\frac{\Delta y}{\Delta x}\right)\left(\frac{\Delta x}{\Delta t}\right) \approx \left(\frac{dy}{dx}\right)\left(\frac{dx}{dt}\right).$$

We now have something to prove that can be sorted out by appealing to first principles, and the definition of "limit."

The general case is sorted out in exactly the same way. And so we get the result anticipated.

Theorem 1.2 (The chain rule in one dimension).
If $y = f(x)$, $x = x(t)$, and $y(t) := y(x(t))$, then

$$\left(\frac{dy}{dt}\right) = \left(\frac{dy}{dx}\right)\left(\frac{dx}{dt}\right).$$

Exercises

Exercise 1.9.1

A car gets 30 miles per gallon on the highway. On a particular trip, the car is using up 2 gallons per hour. From the discussion in the text, the speed of the car is 60 m.p.h. Use the chain rule to find the same result.

Exercise 1.9.2

For $y = \sqrt{3x + 4}$ and $x = (t + 1)^2$, find $\dfrac{dy}{dt}$.

Exercise 1.9.3

For $y = \dfrac{x}{x + 2}$ and $x = t + \sin t$, find $\dfrac{dy}{dt}$.

Exercise 1.9.4

For $y = \displaystyle\int_0^{\cos x} \sqrt{1 + t^4}\, dt$, find $\dfrac{dy}{dx}$.

Exercise 1.9.5

For $y = \displaystyle\int_x^{x^2} \sqrt{1 + t^3}\, dt$, find $\dfrac{dy}{dx}$.

1.10 An elementary example of ratios in integration in one variable

Before getting into the subtleties of applying the chain rule in integration, let's explore an example where integration is accomplished through a single change of scale. The example may at first seem overly elementary. But it carries many aspects of the problem. At the same time, if one is to understand the chain rule and its applications to integration, one needs to be able to explain how this all plays out in even the simplest of examples.

Example

A carpenter who builds furniture buys wood from a local mill. The mill deals with large quantities and sells by the foot. The "2 by 2" (lengths of wood whose cross-section is $2'' \times 2''$) is selling at a price of $6\dfrac{\$}{\text{foot}}$. (Note that we write the dollar sign '\$' after the number 6. This notation is non-standard in banking and finance. However, writing it that way better fits the mathematical symbolism.)

Figure 1.22   Lengths of $2'' \times 2''$ wood sell at 6 \$ per foot.

As it happens, units that are more convenient for the carpenter (both for building furniture and for accounting) are $\dfrac{\$}{\text{inch}}$. In those units, the selling price of the wood is

$$\left(\frac{6\$}{\text{foot}} \times \frac{1 \text{ foot}}{12 \text{ inches}} \right) = \left(\frac{1}{2} \right) \left(\frac{\$}{\text{inch}} \right).$$

Question: The carpenter needs to buy $120''$ of wood. How do we compute the cost?

Again, getting the answer is not a problem. For instance, since we now have the cost per inch, we could simply multiply

$$(120 \text{ inches}) \times \left[\left(\frac{1}{2} \right) \left(\frac{\$}{\text{inch}} \right) \right] = 60\$.$$

Or, we could convert the length of 120″ into feet. Then,

$$10(\text{feet}) \times \left(6 \left(\frac{\$}{\text{foot}} \right) \right) = 60\$.$$

Not surprisingly, the answer is the same in both cases. For, whether we use inches or feet, it is the same length of wood and the cost is given in the same currency.

The problem posed here, however, is not just to find the answer but to understand how the answer is reached. Given the cost in units of

$$\frac{\$}{\text{foot}},$$

and given that we need to purchase so many inches of wood (e.g., 120″), how do we re-express all of that in inches and

$$\frac{\$}{\text{inch}}?$$

An elementary solution for the case when ratios are constant

Whether we use feet or inches, the cost per unit length is constant and so, just as in elementary school, the problem can be solved by a strategic product of fractions.

The systematic solution for converting is:

Cost in $

$$= [10(\text{feet})] \times \left[6 \left(\frac{\$}{\text{foot}} \right) \right]$$

$$= \left[10(\text{feet}) \left(\frac{12''}{\text{foot}} \right) \right] \times \left[6 \left(\frac{\$}{\text{foot}} \left(\frac{1 \text{ foot}}{12 \text{ inches}} \right) \right) \right]$$

$$= (120 \text{ inches}) \left[\left(\frac{1}{2} \right) \left(\frac{\$}{\text{inch}} \right) \right] = 60\$.$$

Notice in the second (middle) expression that, because of introducing inches, the "number" for the length is multiplied a factor of 12 while the number for increments is multiplied by a factor of $\frac{1}{12}$. In other words, the two multipliers cancel, leaving the cost the same.

Cases where ratios and scales are non-constant

The problem is to *add*, to get a total cost for a length of wood. There are *lengths* (intervals) and (as will be shown) there is a *cost density function* $C(x)$ defined on the interval.

In the example, just discussed that cost density function happens to be constant. And so the change of scale from feet to inches is the same change of scale all along. But when buying timber, cost density functions need not be constant.

For instance, for the construction of a specially designed A-frame hall, the cross-section of large wooden beams may need to gradually expand toward their base. This is to meet both design and engineering requirements. In this scenario, wider cross-sections have more volume and so cost more.

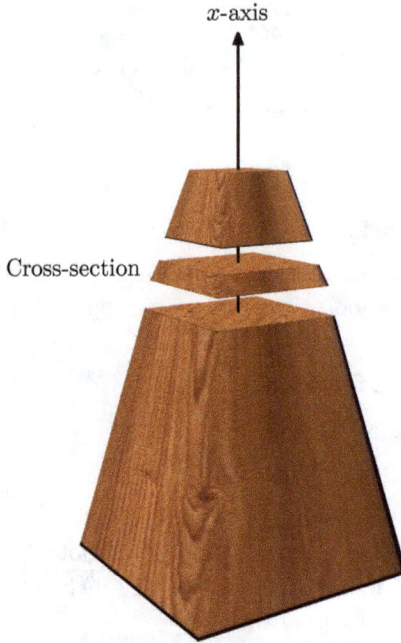

Figure 1.23   Tapered beam of wood. Cost of cross-section depends on width.

Or again, because the range of forces is so large in earthquakes, to measure intensity it is convenient to use exponents to distinguish cases. Measuring devices are calibrated to provide not *Newtons* but a weighted logarithm of Newtons known as the *Richter scale*. In fact, logarithmic scales are used in many applications, including in the study of sound frequencies, for the which a standard scale is *decibels*.

Our task is to be able to handle all such cases, that is, where $C(x)$ is not necessarily constant; and also where scale may vary not by some constant ratio but according to some function $x = x(t)$.

The key is that for, for small intervals, we can use linear approximation and then add results. And so we can expect that the general solution will be an integral. However, in anticipation of handling the general case, let's first return to the example where the carpenter is buying wood.

Temporary return to the elementary carpentry example

Let's treat each aspect of the problem in turn. It is a matter of writing the elementary solution already given, but now with an eye for parts of the solution that in the general case might vary.

Let x be feet. For the mill, the cost density function near x is approximated by

$$\frac{\Delta\$}{\Delta(\text{feet})}.$$

In this example, for a small segment, this ratio is

$$\frac{(\Delta x \text{ feet}) \times \dfrac{6\$}{\text{foot}}}{(\Delta x \text{ feet})}.$$

In other words, as we already knew, the cost-density function is the constant function

$$C(x) = 6\frac{\$}{\text{foot}}.$$

The length of wood to be purchased is 10 feet.

Partitioning the length of wood into increments of equal length Δx, the total cost is

$$\text{Total Cost} = \lim_{\Delta x \to 0} \sum_{i=1}^{n} \left\{ \left[C(x_i) \left(\frac{\$}{\text{foot}} \right) \right] [\Delta x \ (\text{feet})] \right\}.$$

Or, in Leibniz notation,

$$\text{Total Cost} = \int_{x=0}^{x=10} C(x)\, dx = \int_{x=0}^{x=10} 6\, dx = 6\left(\frac{\$}{\text{foot}} \right) [(10 - 0)(\text{feet})] = 60\$.$$

To meet the needs of the carpenter, we can redo this by changing the length scale from feet to inches.

Let $t = $ inches. The relationship between inches and feet (between x and t) is $t = 12x$.

In order to calculate the integral in the new scale, we need to first convert all terms of the integral to the new scale.

(1) The new interval is $t(0) = 12(0) = 0$ and $t(10) = 12(10) = 120$.

(2) The function $C(x)$ is constant and so, numerically, $C(x(t)) = 6$ and units are $\left(\dfrac{\$}{\text{foot}}\right)$. The units here are not right. Remember, we are converting to inches. However, there is no cause for concern. Much as in the solution to the elementary case, once we put everything together, units will "self-adjust."

(3) Since t (inches) $= 12x$ (feet), an increment dt in inches correlates with an increment dx in feet by the relation $dt = 12dx$. In other words,

$$dx = \left[\left(\frac{1}{12}\right)dt\right].$$

Substituting into the integral that was given in terms of feet, and tracking units,

$$\text{Total Cost} = \int_{x=0}^{x=10} C(x)dx = \int_{x=0}^{x=10}\left[6\left(\frac{\$}{\text{foot}}\right)\right][dx(\text{feet})]$$

$$= \int_{t=0}^{t=120}\left[6\left(\frac{\$}{\text{foot}}\right)\right]\left(\frac{1}{12}\frac{\text{foot}}{\text{inch}}\right)[dt(\text{inches})].$$

If we suppress explicit identification of units, the algebraic structure is more visible. That is,

$$\text{Total Cost} = \int_{x=0}^{x=10} C(x)dx = \int_{x=0}^{x=10} 6dx$$

$$= \int_{t=1}^{t=120} 6\left(\frac{1}{12}dt\right) = (120)\left(\frac{6}{12}\right) = 60.$$

In other words, from (1), the change of variables leads to a change in the interval. From (2), we see what the cost function looks like in the new variable. From (3), a multiplying factor compensates for the change in (1). Observe, also, that the net effect of the (internal) change of variables is that the $-value of the integral is invariant.

1.11   The general case: applications of the chain rule in integration in one variable

The main idea is that by reversing the chain rule we get an integration formula.

Example

Consider the integral $\displaystyle\int_1^2 3(t^5 + t)^2(5t^4 + 1)dt$.

One may spot that the integrand can be expressed as a derivative of a composition.

By the chain rule,

$$3(t^5 + t)^2(5t^4 + 1) = \frac{d}{dt}(t^5 + t)^3$$

and

$$\int_1^2 3(t^5 + t)^2(5t^4 + 1)dt = \int_1^2 \frac{d}{dt}(t^5 + t)^3 dt$$

By the Fundamental Theorem of Calculus, this last integral is

$$\int_1^2 \frac{d}{dt}(t^5 + t)^3 dt = (t^5 + t)^3\Big|_1^2 = (34)^3 - (2)^3.$$

What if an anti-derivative is not quite so obvious? One approach is called the "**method of substitution**":

That is, we use Leibniz notation for the chain rule and compute accordingly: if $y = y(x)$ and $x = x(t)$ then

$$\frac{dy}{dt} = \frac{dy}{dx}\frac{dx}{dt}.$$

In the example just given, on a hunch that the integrand might be brought under control, let $x = t^5 + t$. Then,

$$\frac{dx}{dt} = (5t^4 + 1).$$

Leibniz notation is then convenient for expressing the corresponding linear approximation:

$$dx = (5t^4 + 1)dt.$$

By substitution, the integral can be expressed as

$$\int_c^d 3(x)^2 dx,$$

where the interval of integration is yet to be determined.

For the interval of integration, observe that when $t = 1$, $x = 2$ and when $t = 2$, $x = 34$. Hence, the integral is

$$\int_2^{34} 3(x)^2 dx = x^3 \Big|_2^{34} = (34)^3 - (2)^3.$$

The argument in the general case is the same. By the chain rule,

$$\int_c^d \left(\frac{df}{dx}\right) \left(\frac{dx}{dt}\right) = \int_c^d \frac{d}{dt}\Big(f(x(t))\Big) dt.$$

By the Fundamental Theorem of Calculus in one variable,

$$\int_c^d \frac{d}{dt}\Big(f(x(t))\Big) dt = f\Big(x(d)\Big) - f\Big(x(c)\Big)$$

Hence, the integral

$$\int_c^d \frac{d}{dt}\Big(f(x(t))\Big) dt = f\Big(x(d)\Big) - f\Big(x(c)\Big) = f(b) - f(a).$$

As in the example above, the "method of substitution" uses Leibniz notation to compute the new integrand. It is useful when, for instance, a strategic change of variables produces an integrand that is a derivative of the form $\frac{d}{dt}\Big(f(x(t))\Big)$.

To see how this works for an integral

$$\int_a^b y(x)dx,$$

we introduce a change of variables $x = x(t)$.

As in the Example above,

(1) If $x(c) = a$ and $x(d) = b$, then the new interval of integration is $c \le t \le d$.

(2) In terms of the new variable, the new function is obtained by using the normal rules of function composition, that is, $y = y(x) = y(x(t))$.

(3) All of this is an application of the chain rule. In other words, the "method of substitution" is the symbolic technique that takes advantage of the Leibniz abbreviation already used in the development of the chain rule.

As in the example above where $g(t) = 3(t^5+t)^2(5t^4+1)$, we need to compute an integral of the form $\int_c^d g(t)dt$. The strategy is in the hope that a change of variables might bring this into tractable form. We might try some $x(t)$.

By linear approximation, an increment dt along the t axis corresponds to an increment

$$dx = \left[\left(\frac{dx}{dt}\right)dt\right]$$

along the x axis.

If the integrand $g(t)$ is then identified to be of the form

$$y\left(x(t)\right)\left(\frac{dx}{dt}\right),$$

then by the chain rule we get

$$\int_c^d g(t)dt = \int_c^d y\left(x(t)\right)\left(\frac{dx}{dt}\right)dt = \int_{x(c)}^{x(d)} y(x)dx.$$

The new integrand might then be further analyzed. In some cases, such as in the example above, $y(x) = \dfrac{df}{dx}$ for a known function $f(x)$. In that case, by the Fundamental Theorem of Calculus, we conclude with a numerical solution, namely, $f\left(x(d)\right) - f\left(x(c)\right)$.

Exercises

Exercise 1.11.1

Evaluate the definite integral: $\displaystyle\int_0^2 (2x-1)^3\,dx$.

Exercise 1.11.2

Evaluate the definite integral: $\displaystyle\int_1^3 \frac{x}{x^2+3}\,dx$.

Exercise 1.11.3

Evaluate the definite integral: $\displaystyle\int_0^\pi \sin 3x\,dx$.

Exercise 1.11.4

Evaluate the definite integral: $\int_0^{\pi/4} \tan x \, dx$.

In Exercises 1.11.5–1.11.6, find the given area.

Exercise 1.11.5

The area of the region bounded by $y = \dfrac{\ln x}{x}$ and the x-axis for $1 \leq x \leq e$.

Exercise 1.11.6

The area of the region bounded by $y = \dfrac{4x}{x^2 + 1}$ and the x-axis for $0 \leq x \leq 1$.

1.12   The chain rule in two dimensions and one independent variable

Suppose that $z = f(x, y)$ and $(x(t), y(t))$ is a curve in the (x, y) plane.

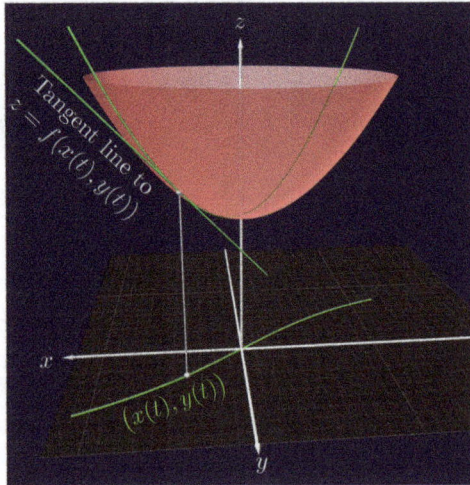

Figure 1.24   The curve $z(t) = f(x(t), y(t))$ has a tangent line and slope at $(x(t), y(t))$.

The problem is to determine the slope of the tangent line of the curve $z = f(x(t), y(t))$ in (x, y, z) space. In other words, what is $\dfrac{dz}{dt}$ where $z = f(x, y)$ and each of x and y are functions of t, $x = x(t)$ and $y = y(t)$?

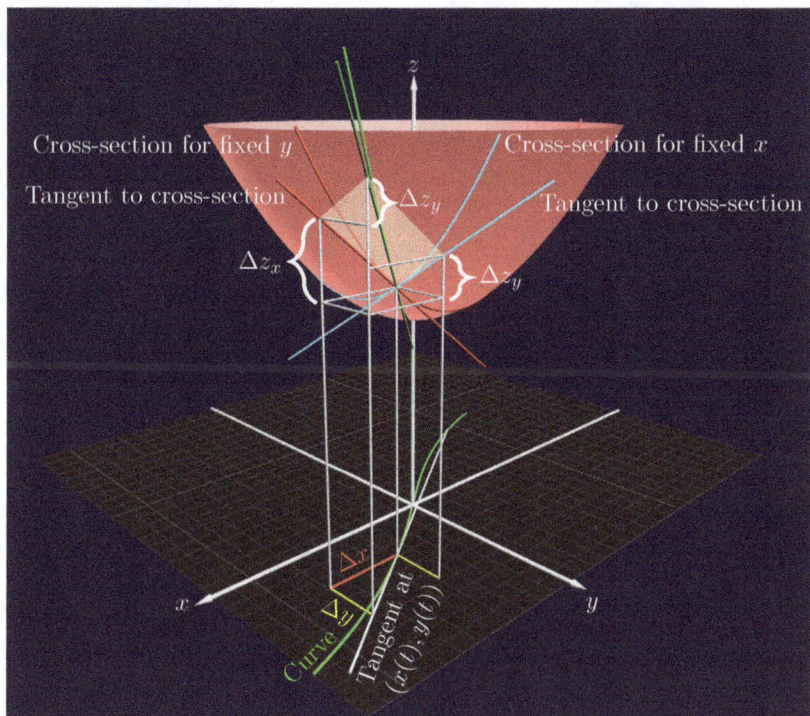

Figure 1.25   To compute net change, use Δx, Δy and partial derivatives of $z = f(x, y)$ to compute $\Delta z_{x\text{-direction}}$ and $\Delta z_{y\text{-direction}}$. Then add.

Using the one-dimensional result together with the diagram, tracking change in one direction at a time, the result follows:

(1) For a small increment Δt, there are small increments

$$\Delta x = \left(\frac{dx}{dt} \right) \Delta t$$

and

$$\Delta y = \left(\frac{dy}{dt} \right) \Delta t.$$

(2) For the small increments in (1), there are two small increments in $z = f(x, y)$:

$$(\Delta z)_{\text{in } x \text{ direction}} = \frac{\partial f}{\partial x} \Delta x$$

and

$$(\Delta z)_{\text{in } y \text{ direction}} = \frac{\partial f}{\partial y} \Delta y.$$

The total increment Δz in $z = f(x, y)$ is, then, approximately

$$\Delta z = (\Delta z)_{\text{in } x \text{ direction}} + (\Delta z)_{\text{in } y \text{ direction}}$$

$$= \frac{\partial f}{\partial x} \Delta x + \frac{\partial f}{\partial y} \Delta y$$

$$= \frac{\partial f}{\partial x} \left(\frac{dx}{dt} \right) \Delta t + \frac{\partial f}{\partial y} \left(\frac{dy}{dt} \right) \Delta t$$

Dividing by Δt, and taking the limit, the result follows. That is, if $z = f(x(t), y(t))$, then

$$\frac{dz}{dt} = \frac{\partial z}{\partial x} \frac{dx}{dt} + \frac{\partial z}{\partial y} \frac{dy}{dt}.$$

Exercises

In Exercises 1.12.1–1.12.3, use the Chain Rule to find $\dfrac{dz}{dt}$.

Exercise 1.12.1

$z = (x + y)^2 + xy$, $x = 2\sqrt{t}$, $y = t^3$

Exercise 1.12.2

$z = \sin x \cos y$, $x = 3t$, $y = t + 4$

Exercise 1.12.3

$z = \ln (x^2 + y)$, $x = e^{-3t}$, $y = te^{3t}$

Exercise 1.12.4

Suppose that a curve in space tracks along the surface of the graph of $z = x^2 + y^2$, and that the projection of the curve onto the x-y plane is given by the curve $\mathbf{r}(t) = (\cos t, \sin t, 0)$. Find the slope of the tangent line to the curve in three dimensions. Graph your results and explain.

Exercise 1.12.5

Suppose that a curve in space tracks along the surface of the graph of $z = x^2 + y^2$, and that the projection of the curve onto the x-y plane is given by the curve $\mathbf{r}(t) = (4 \cos t, 3 \sin t, 0)$. Find the slope of the tangent line to the curve in three dimensions. Graph your results and explain.

1.13 The chain rule in two dimensions and two independent variables

In mathematics and its applications, it is often useful to be able to convert results from one coordinate system to another. A familiar example is between (r, θ) polar coordinates and (x, y) rectangular coordinates. But the problem also comes up in maps and atlases. For instance, locations given relative to coordinates lined up with a local river system and watershed might need to be translated to a map of a larger region based on a north-south and east-west grid.

Example

Let's begin by looking to the familiar case of polar coordinates and rectangular coordinates. Both sets of coordinates can be used to determine all points of the Euclidean plane. The plane is endowed with the Pythagorean distance function.

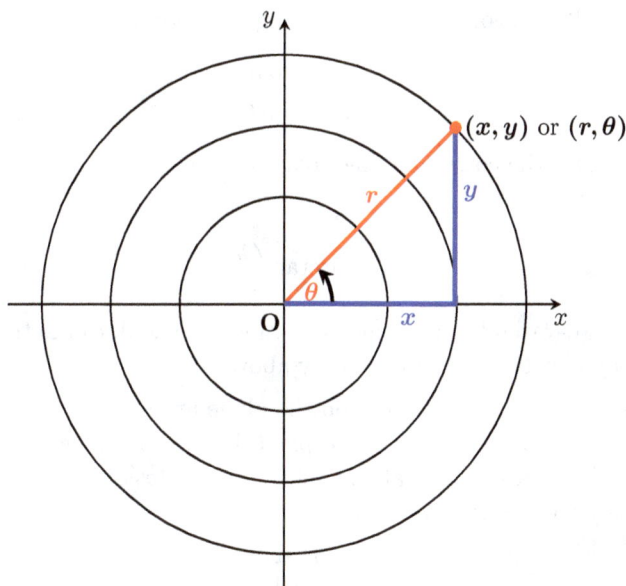

Figure 1.26   Polar and Cartesian coordinates for Euclidean plane.

A first set of coordinate curves consists of mutually perpendicular lines, providing what is known as (x, y) coordinate lines. An origin O is selected.

Centered at the selected origin O, another set of coordinate curves can be introduced, namely, radial arms emanating from the origin O together with circles centered at the origin O.

For present purposes, we can restrict our attention to points that are interior to the first quadrant of the (x, y) plane, that is, where $x > 0$ and $y > 0$.

Using these two different sets of coordinate curves, each point P on the plane can be represented in two different ways:

(1) by a pair (x, y), where x and y refer to distances along rectangular coordinate lines relative to the selected origin; and

(2) by a pair (r, θ) , where r is the distance from the origin to the P and θ is the angle (in radians) from the line $y = 0$ to the radial line containing P.

There are translation equations going both ways:

(1) Given (r, θ), the corresponding rectangular coordinates are

$$x = r \cos \theta$$
$$y = r \sin \theta.$$

(2) Given (x, y), the corresponding polar coordinates are

$$r = \sqrt{x^2 + y^2}$$
$$\theta = \arctan \left(\frac{y}{x} \right).$$

To see the "two-variable chain rule" in action, let's look to the transformation given by equation (1) immediately above.

Suppose that two points P and Q on the plane are relatively close to each other. (We can do this because the plane has the Pythagorean distance function.) We suppose also that the polar coordinates of both P and Q are known, as are the increments dr and $d\theta$.

As in examples of the chain rule given above, the problem is to determine the corresponding increments in the variables (x, y). This is a valid question because $x = x(r, \theta)$ and $y = y(r, \theta)$.

Given $P = P(x, y) = P(r, \theta)$ and $Q = Q(r + dr, \theta + d\theta)$, what are the corresponding increments dx and dy by which to (linearly) approximate $Q = Q(x + dx, y + dy)$ in rectangular coordinates?

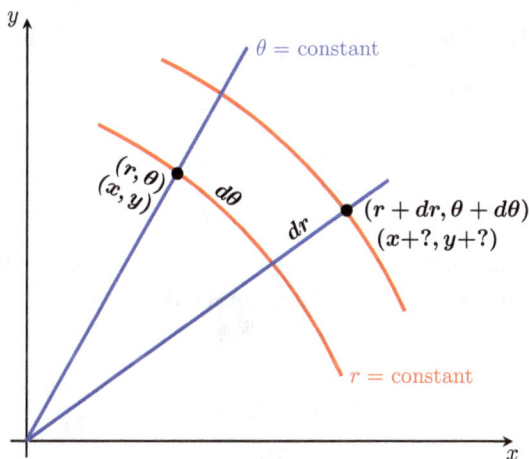

Figure 1.27 Given (r, θ) and (x, y), $x = r \cos \theta$, $y = r \sin \theta$ and $(r + dr, \theta + d\theta)$, what is the new point in Cartesian coordinates?

Example

Let's start with the x coordinate. From P to Q there is a change in both r and θ, namely, dr and $d\theta$. As described in previous examples, through linear approximation the increment in x is a sum of an increment determined by a change in r and an increment determined by a change in θ. That is,

$$dx = \frac{\partial x}{\partial r} dr + \frac{\partial x}{\partial \theta} d\theta. \tag{1.3}$$

In the same way,

$$dy = \frac{\partial y}{\partial r} dr + \frac{\partial y}{\partial \theta} d\theta. \tag{1.4}$$

Since points in the plane are determined by pairs of coordinates, this is written more compactly, as follows:

For a pair of increments in polar coordinates dr and $d\theta$, the corresponding increments in rectangular coordinates are given by

$$\begin{pmatrix} dx \\ dy \end{pmatrix} = \begin{pmatrix} \dfrac{\partial x}{\partial r} & \dfrac{\partial x}{\partial \theta} \\ \dfrac{\partial y}{\partial r} & \dfrac{\partial y}{\partial \theta} \end{pmatrix} \begin{pmatrix} dr \\ d\theta \end{pmatrix} = \frac{\partial(x, y)}{\partial(r, \theta)} \begin{pmatrix} dr \\ d\theta \end{pmatrix}.$$

Definition 1.1. The matrix $\dfrac{\partial(x,y)}{\partial(r,\theta)}$ is an abbreviation for

$$\begin{pmatrix} \dfrac{\partial x}{\partial r} & \dfrac{\partial x}{\partial \theta} \\[2ex] \dfrac{\partial y}{\partial r} & \dfrac{\partial y}{\partial \theta} \end{pmatrix},$$

the first-order linear approximation of relative rates of change and is also called the *Jacobian* matrix for equation (1).

Problem 1.8.

Show that in the case of polar coordinates,

$$\frac{\partial(x,y)}{\partial(r,\theta)} = \begin{pmatrix} \cos\theta & -r\sin\theta \\ \sin\theta & r\cos\theta \end{pmatrix}.$$

The general case

To get the general result, observe that the computations above did not depend on the fact that r and θ were polar coordinates. The key is that there is a functional dependence of the form $x = x(r,\theta)$ and $y = y(r,\theta)$.

For the general case, we write $x = x(u,v)$ and $y = y(u,v)$; and represent coordinates (u,v) by their own coordinate plane.

The problem is: Given the pair of increments $\begin{pmatrix} du \\ dv \end{pmatrix}$, what is the corresponding pair of increments $\begin{pmatrix} dx \\ dy \end{pmatrix}$?

As with the special case of polar coordinates, we use linear approximation given by the Jacobian matrix

$$\begin{pmatrix} \dfrac{\partial x}{\partial u} & \dfrac{\partial x}{\partial v} \\[2ex] \dfrac{\partial y}{\partial u} & \dfrac{\partial y}{\partial v} \end{pmatrix}.$$

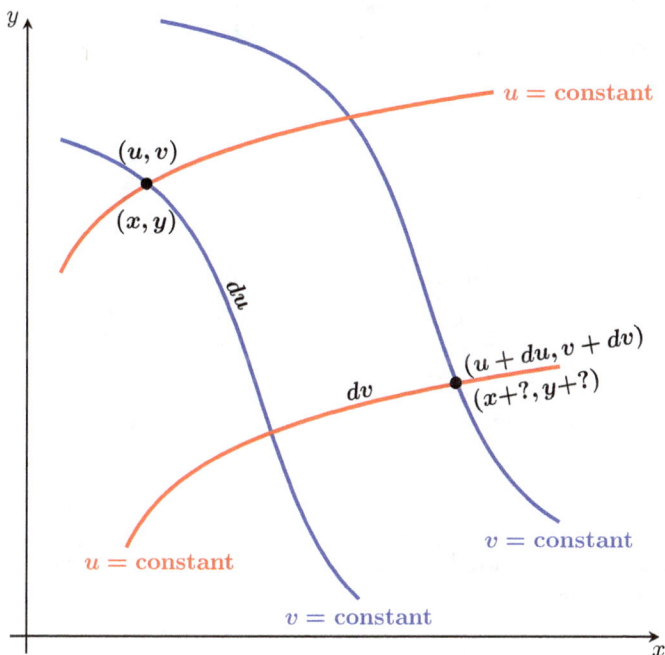

Figure 1.28 Given (u, v) and (x, y), $x = x(u, v)$, $y = y(u, v)$ and $(u + du, v + dv)$, what is the new point in Cartesian coordinates?

The chain rule for two dimensions and two independent variables is

$$\begin{pmatrix} dx \\ dy \end{pmatrix} = \begin{pmatrix} \dfrac{\partial x}{\partial u} & \dfrac{\partial x}{\partial v} \\[2ex] \dfrac{\partial y}{\partial u} & \dfrac{\partial y}{\partial v} \end{pmatrix} \begin{pmatrix} du \\ dv \end{pmatrix} = \frac{\partial(x, y)}{\partial(u, v)} \begin{pmatrix} du \\ dv \end{pmatrix}.$$

Exercises

Exercise 1.13.1

Show that the mapping $(x, y) \longrightarrow (r, \theta)$ is one-one and onto.

In Exercises 1.13.2–1.13.4, use the Chain Rule to find $\dfrac{\partial z}{\partial s}$ and $\dfrac{\partial z}{\partial t}$.

Exercise 1.13.2

$z = x^3 + xy$, $x = s + t$, $y = st$

Exercise 1.13.3

$z = x + xy + 4$, $x = s/t$, $y = s - 3t$

Exercise 1.13.4

$z = ye^{x/y}$, $x = st$, $y = t/s$

In Exercises 1.13.5–1.13.6, find the Jacobian matrix $\dfrac{\partial(x, y)}{\partial(u, v)}$.

Exercise 1.13.5

$x = 2u - 3v$ and $y = 5u + 7v$

Exercise 1.13.6

$x = 3u + uv$ and $y = \dfrac{v}{u^2 + v^2}$

1.14 Implications for integration over a region in two variables

Variables are as stated in Section 1.13, in the general case. As in the one-variable chain rule, there are applications in two-variable integration.

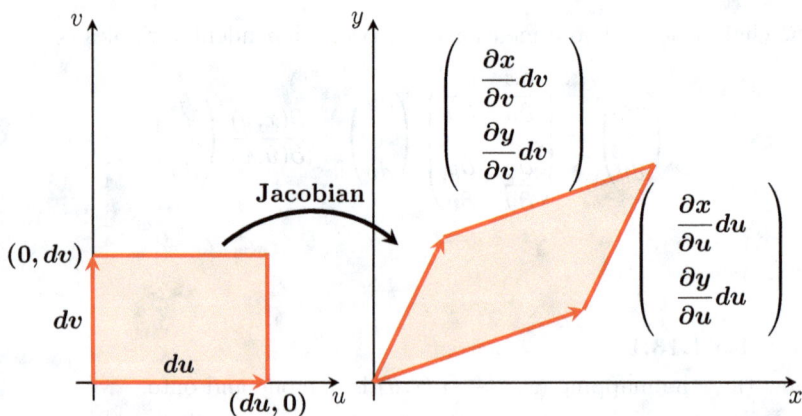

Figure 1.29   The linear approximation maps a rectangle in (u, v) plane to a parallelogram in (x, y) plane.

In (u, v) coordinates, an element of area is of the form $du \times dv$. This is a small rectangle, one side of which is the vector

$$\begin{pmatrix} du \\ 0 \end{pmatrix}$$

and the other side of which is the vector

$$\begin{pmatrix} 0 \\ dv \end{pmatrix}.$$

What is the corresponding area element measured in the (x, y) coordinate system?

Problem 1.9.

By linear approximation to the transformation

$$x = x(u, v) \text{ and } y = y(u, v)$$

and the chain rule, the rectangle

$$du \times dv$$

is linearly mapped (by the Jacobian) to a parallelogram in (x, y) coordinates, with sides

$$\begin{pmatrix} \dfrac{\partial x}{\partial u} du \\[2mm] \dfrac{\partial y}{\partial u} du \end{pmatrix} \text{ and } \begin{pmatrix} \dfrac{\partial x}{\partial v} dv \\[2mm] \dfrac{\partial y}{\partial v} dv \end{pmatrix}.$$

From our previous work on area functions for the Euclidean plane, the area of this parallelogram is

$$\left| \det \begin{pmatrix} \dfrac{\partial x}{\partial u} du & \dfrac{\partial x}{\partial v} dv \\[2mm] \dfrac{\partial y}{\partial u} du & \dfrac{\partial y}{\partial v} dv \end{pmatrix} \right| = \left| \left(\det \frac{\partial(x, y)}{\partial(u, v)} \right) du dv \right| = \left| \left(\frac{\partial x}{\partial u} \frac{\partial y}{\partial v} - \frac{\partial y}{\partial u} \frac{\partial x}{\partial v} \right) du dv \right|.$$

This provides the area element and we get the well-known "change of variable" formula for integrals:

If the region $S(u, v)$ in (u, v) corresponds to $R(x, y)$ in (x, y), then

$$\iint\limits_{R(x,y)} f(x, y) dx dy = \iint\limits_{S(u,v)} f\left(x(u, v), y(u, v) \right) \left| \left(\det \frac{\partial(x, y)}{\partial(u, v)} \right) \right| du dv.$$

Example

In polar coordinates,

$$\det\frac{\partial(x,y)}{\partial(u,v)} = r,$$

and so

$$\iint\limits_{x^2+y^2\leq1} (x^2+y^2)dxdy = \iint\limits_{[0\leq r\leq1]\times[0\leq\theta\leq2\pi]} r^2(r)drd\theta$$

$$= \iint\limits_{[0\leq r\leq1]\times[0\leq\theta\leq2\pi]} r^3 drd\theta$$

$$= (2\pi)\left(\frac{1}{4}\right)$$

$$= \frac{\pi}{2}.$$

Exercises

In the following exercises, draw the coordinate lines $u = u_0$ and $v = v_0$.

Exercise 1.14.1

This exercise concerns the iterated integral

$$\int_0^1 \int_{-y}^y xy\,dx\,dy.$$

(a) Evaluate this integral and sketch the region D of integration in the xy-plane.

(b) Let $u = x + y$ and $v = x - y$. Find the region S in the uv-plane that corresponds to D.

(c) Use the change of variables formula to evaluate the integral by using the substitution $u = x + y$, $v = x - y$.

Exercise 1.14.2

This exercise concerns the iterated integral

$$\int_0^2 \int_y^{8-3y} \cos(x+3y)\sin(x-y)\,dx\,dy.$$

(a) Evaluate this integral and sketch the region D of integration in the xy-plane.

(b) Let $u = x + 3y$ and $v = x - y$. Find the region S in the uv-plane that corresponds to D.

(c) Use the change of variables formula to evaluate the integral by using the substitution $u = x + 3y$, $v = x - y$.

Exercise 1.14.3

Evaluate $\iint\limits_{D}(3x - 2y)dA$, where D is the region enclosed by the ellipse $9x^2 + 16y^2 = 144$. Change the variables using $u = \dfrac{x}{4}$ and $v = \dfrac{y}{3}$.

Exercise 1.14.4

Transform the given integral to one in polar coordinates then evaluate the polar integral.

(a) $\displaystyle\int_{-1}^{1}\int_{-\sqrt{1-x^2}}^{\sqrt{1-x^2}} y^2 \, dy \, dx$

(b) $\displaystyle\int_{-2}^{2}\int_{0}^{\sqrt{4-y^2}} e^{x^2+y^2} \, dx \, dy$

PART 1

FLUID MOTION IN 2 DIMENSIONS

Chapter 2

2-D MASS-FLOW VELOCITY FIELDS

2.1 Surface vector fields for fluid flow

Have you ever watched and wondered about the motion of a stream, or been intrigued by a vortex of water draining from a sink? Perhaps you've watched ground water emerging from a spring (which in French is called "une source").

In applied science programs, water flow patterns are produced and studied with chutes and by mechanically introducing sources and drains (or "sinks").

A relatively simple case is water flowing slowly, at a constant speed in a shallow chute in a laboratory.

A water flow has depth and samples of water have mass that can be measured with weigh scales. There is, however, a problem here that is not encountered in Newtonian physics. In Newtonian physics, one studies motion of an isolated object with a center of mass. This is then "idealized" and mathematized when one speaks of "point-particles" with mass and velocity. By contrast, water flow is a "medium in motion."

However, we can take advantage of classical physics. Introduce a small fragment of cork, or anything that is small and floats along with the water flow. This is sometimes called a "test particle." If we follow a moving test particle, it has velocities in the classical sense. Note also that, in practice, test particles can be introduced at all locations on the surface of a water flow.

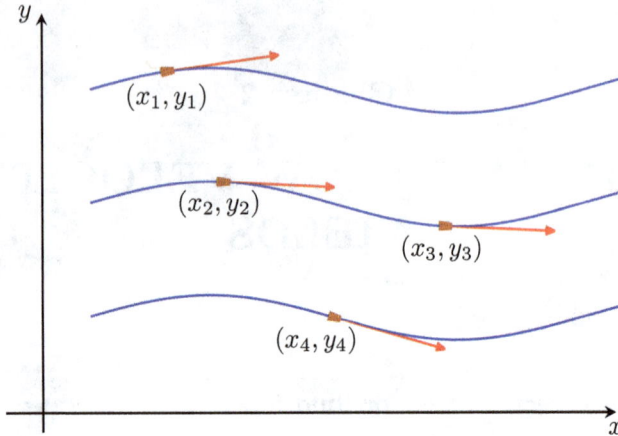

Figure 2.1   Four test particles with velocities at locations (x_1, y_1), (x_2, y_2), (x_3, y_3), (x_4, y_4).

For this book, we are assuming no "turbulence"[26] and that the surface remains (more or less) level. In fact, there are usually waves of at least some amplitude. However, at low speeds, and if measurements need not be extremely refined, then vertical motion can be eliminated from approximations. Note also that by using submerged test particles and transparent walls in a laboratory chute, it is found that velocities below and near the surface are approximately equal to surface velocities. For now, then, we focus only on horizontal components of motion of a shallow water flow.

Note, however, that the velocities of the surface are also understood to represent the velocities of water below and near the surface.

To get started, then, what we need is a 2-component *velocity vector field* for the horizontal velocity of a test particle at location (x, y) on the surface at time t. That is, the velocity vector is of the form

$$\mathbf{v} = \mathbf{v}(x, y, z, t) = \big(a(x, y, z, t), b(x, y, z, t)\big).$$

Remember, however, that velocity is also the velocity of test particles below and near the surface. That is, locally, the velocity is independent of depth z which leads to the notation

$$\mathbf{v} = \mathbf{v}(x, y, t) = \big(a(x, y, t), b(x, y, t)\big).$$

[26] For the present context, we are not providing a definition of *turbulence*. That would be done in a follow-up course in fluid dynamics.

And actual water has breadth and depth.

As in classical physics, we make use of the idealization wherein we speak of "points" that are "locations without breadth or depth."

Problem 2.1.

In this book, we will not be studying the case where there is time-dependence. Instead, we will focus on cases where **v** is of the form

$$\mathbf{v}(x, y) = \big(a(x, y), b(x, y)\big).$$

The velocity field is constant in time t but need not be constant in space coordinates (x, y).

(1) Can you imagine elementary cases where the velocity vector field is independent of time but is not constant in its space coordinates? *Hint*: Imagine a chute or ramp down which water flows due to gravity, and the chute then levelling out further downstream.

(2) Can you produce explicit velocity fields

$$\mathbf{v}(x, y) = \big(a(x, y), b(x, y)\big)$$

to match the phenomena described in (1)? *Hint*: Think of Galileo's result about free-fall.

(3) While we will not be inquiring into the case of time-dependence, for context, it can be helpful to be aware of a few cases. For examples, think of water flow in a chute where the water flow is driven by an oscillator. Or, again, if a chute is engineered to do so, the angle of descent could change in time.

(4) Can you produce explicit vector fields

$$\mathbf{v}(x, y) = \big(a(x, y), b(x, y)\big)$$

to match the phenomena described in (3)?

Problem 2.2.

By using small time increments and first-order terms, work out that speed of a surface near (x, y) is given by

$$\| \mathbf{v} \| = \sqrt{a^2 + b^2}.$$

Hint: With projections, treat one component at a time and use the Euclidean distance function on the 2-dimensional surface.

Exercises

In Exercises 2.1.1–2.1.10, graph the velocity field, for $x \geq 0$. Try this without using a graphing calculator or graphing program. Determine the speed and describe the fluid motion.

Exercise 2.1.1

$\mathbf{v}(x, y) = (3, 0)$

Exercise 2.1.6

$\mathbf{v}(x, y) = (3x + y, x + y)$

Exercise 2.1.2

$\mathbf{v}(x, y) = (x, 0)$

Exercise 2.1.7

$\mathbf{v}(x, y) = (1, \cos x)$

Exercise 2.1.3

$\mathbf{v}(x, y) = (3x, x)$

Exercise 2.1.8

$\mathbf{v}(x, y) = (x, \cos x)$

Exercise 2.1.4

$\mathbf{v}(x, y) = (x, y)$

Exercise 2.1.9

$\mathbf{v}(x, y) = (\cos^2 x, \sin x)$

Exercise 2.1.5

$\mathbf{v}(x, y) = (-y, x)$

Exercise 2.1.10

$\mathbf{v}(x, y) = (\sin^2 x, \cos^2 x)$

2.2 Water flows, streamlines and integral curves

Test particles follow *trajectories* called *streamlines*. The velocity vector of a test particle is tangent to the streamline. On the other hand, given a velocity field $\mathbf{v}$ and an initial point, a curve $\mathbf{r}(u) = (x(u), y(u))$ is a streamline or integral curve when at every point along the curve its velocity vector is $\mathbf{v}$. That is, we require

$$\frac{d}{du}(\mathbf{r}(u)) = (x'(u), y'(u)) = \mathbf{v}(\mathbf{r}(u)).$$

For details, see a Calculus III textbook.

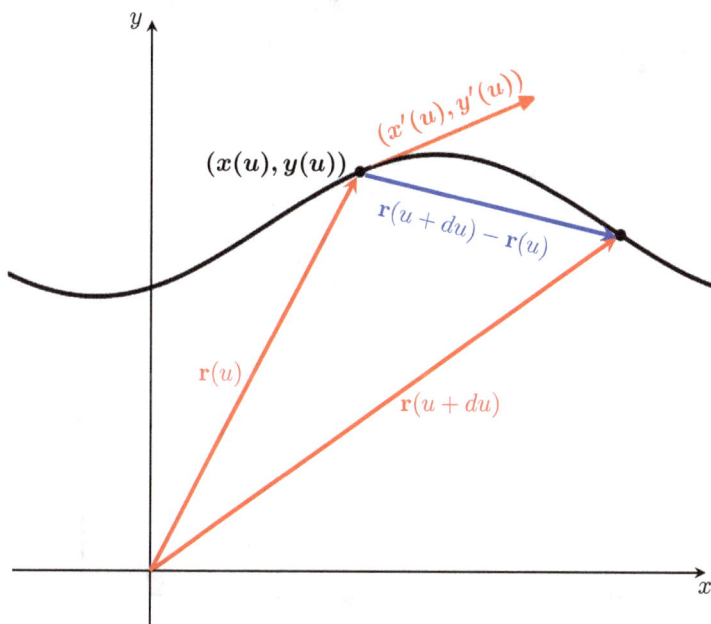

Figure 2.2   Velocity vector $(x'(u), y'(u))$ of a parameterized streamline. $\mathbf{v}(x(u), y(u)) = \lim_{du \to 0} \dfrac{r(u+du) - r(u)}{du} = (x'(u), y'(u))$.

Exercises

In Exercises 2.2.1–2.2.8, compute the integral curve for the given velocity field.

Exercise 2.2.1

$\mathbf{v}(x, y) = (3, 0)$

Exercise 2.2.4

$\mathbf{v}(x, y) = (x, y)$

Exercise 2.2.2

$\mathbf{v}(x, y) = (x, 0)$

Exercise 2.2.5

$\mathbf{v}(x, y) = (-y, x)$

Exercise 2.2.3

$\mathbf{v}(x, y) = (3x, x)$

Exercise 2.2.6

$\mathbf{v}(x, y) = (3x + y, x + y)$

Exercise 2.2.7

$\mathbf{v}(x, y) = (1, \cos x)$

Exercise 2.2.8

$\mathbf{v}(x, y) = (x, \cos x)$

2.3   Mass flow rates

Imagine a 1 *cm* cross-section (in the direction of the *y*-axis) of a shallow water flow moving in the *x*-direction with a constant speed of, say, 10 *cm* per second.

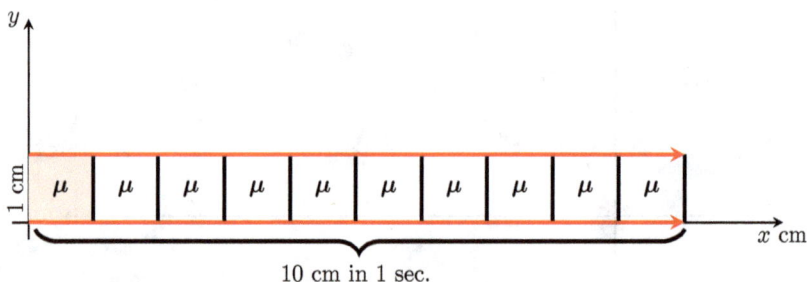

Figure 2.3   Velocity with *x*-component = 10 cm/sec and linear mass density μ g/cm. Total mass-flow $10 \cdot \mu$ g in 1 second.

Keep in mind that our inquiry begins with water and other fluid flows. And so, we can take advantage of what is known in elementary physics. A key quantity in physics is *momentum*, namely, "(mass)×(velocity)." Classical physics investigates the motion of objects with centers of mass, but we are investigating water flow. We can adapt the definition of *momentum* to water flow.

In the situation just described, what is the rate at which mass flows in the *x*-direction across a 1 *cm* cross-section parallel to the *y*-axis? In other words, how many grams of water flow in the *x*-direction, per second, when the velocity of the flow is 10 *cm* per second in the *x*-direction and the cross-section (*y* direction) is 1 *cm*?

In one second, $10 \cdot 1$ square *cms* of water pass the 1 *cm* cross-section. In a water flow, below each square *cm* of the surface is a volume of water with so many grams per cubic *cm*. Remember that, in our notation, we have suppressed the depth component, but depth is there. This means that we can introduce a linear mass density term, μ. In other words, μ represents the number of grams per linear *cm*. This can be defined and measured because, while the vector field has only two components, we are referring to volumes of water in 3 dimensions, and volumes have mass.

In our example, then, the mass-flow per second that passes a 1 *cm* cross-section is

$$\underbrace{(\mu \cdot 1) + (\mu \cdot 1) + \cdots + (\mu \cdot 1)}_{10 \text{ terms}} = 10 \cdot \mu$$

2.4 Units for mass flow rates

Let's pause here to keep track of units. This will be crucial in applications as well as in developing mathematical results.

The term μ is "grams per *cm* in the *x*-direction"; *speed* is "*cms* per second"; and *length* is "*cm* in the *x*-direction". So, the units of '$10 \cdot \mu$' are "$(cm/sec) \cdot (grams/cm)$". In other words, the quantity $10 \cdot \mu$ is a *rate*, "grams per second" (with the understanding that rate is of a 3-d flow of water moving in the *x*-direction).

Problem 2.3.
In fluid dynamics, μ is a function of both space and time coordinates. For the present context, however, we suppose that μ is constant. We have been talking about water flow. For water, μ is approximately $1\,g/cm$, since the *gram* was originally defined as the *mass of one cubic centimetre of water at its maximum density at* $4°C$. However, our inquiry here anticipates the study of other fluids too, for which μ is not unity. Nevertheless, with constant μ, it is with no loss of generality that we develop our results for $\mu = 1$. Why not? If, for instance, $\mu = 8$, how might we adjust computations so that a new μ, μ' say, is unity? *Hint: volume* $= (length)^3$. Might we replace *cm* with a new scale? If so, what might that new scale be? How should it be related to *cm*?

Problem 2.4.
If time is some fraction of a second, if the cross-section in the *y*-direction is some fraction of a *cm*, or if the depth is something different than 1 *cm* (see hypotheses given in the second paragraph, above), then mass-flow rates need to be adjusted. Explore different cases. What are the appropriate formulas?

Chapter 3

CIRCULATION AND GREEN'S THEOREM

3.1 Circulation (mass flow rate) along a parallel line segment

In studying fluid flow, an engineer may need to know the total mass flow rate, not only near an initial reference line (like the y-axis), but along a length. Suppose the flow and cross-section are as described above. Extend your view to include a 5-cm length in the positive x-direction.

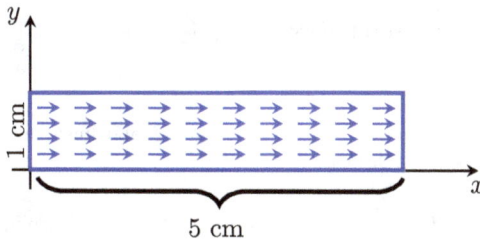

Figure 3.1   Water flow along a $5\,cm \times 1\,cm$ area.

Note that, numerically, the surface area of a 5×1 surface also is 5, but units are cm^2.

If we first look at the $1 \times 1\,cm^2$ immediately adjacent to the y-axis in the x-direction, then in one second, $10\,cm$ flows through that square. But the segment along the x-axis that we are imagining is $5\,cm$ in length and $1\,cm$ wide.

Imagine the flow originating well to left (from somewhere in the negative x-direction) and continuing well to the right (in the positive x-direction).

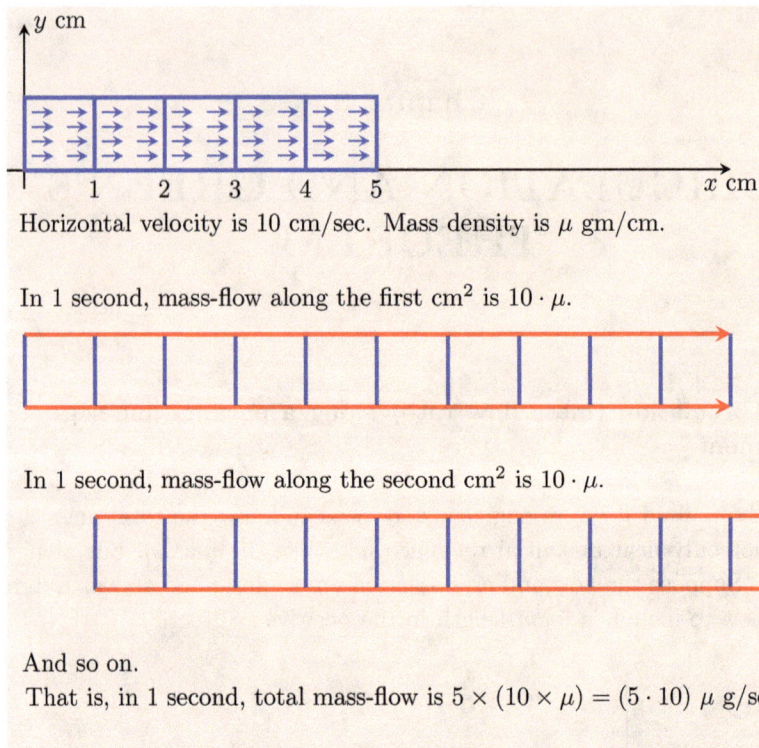

Figure 3.2 Velocity with *x*-component = 10 cm/sec and linear mass density μ g/cm. Total mass-flow $10 \cdot \mu$ g in 1 second.

Through each of the five 1×1 cm^2 along that length, there is a flow rate of 10 centimeters per second. So, with a 1-*cm* cross section, the total *mass-flow rate* through the entire length in the *x*-direction is

$$(10\mu) + (10\mu) + (10\mu) + (10\mu) + (10\mu) = (5 \cdot 10)\mu \text{ grams per second.}$$

This *rate* is called the *circulation (of mass) along the 5-cm length.*

(Note that we are not replacing $5 \cdot 10$ by 50. This is intentional, in order to help us keep track of units and sources of terms. This will be helpful as we go on to other cases.)

Problem 3.1.

In the example described, what is the total mass in motion, over a 3 second *time* period?

The purpose of Problem 3.1 is to emphasize that units for circulation are rates, a total of "grams per second along a curve." If an engineer needs to determine not mass flow rate along a curve, but total mass in motion along a curve over a period of time, then they would need to also compute (or approximate) a time integral.

3.2 Circulation (mass flow rate) of a constant velocity field along an arbitrary line segment in 2-d

Suppose velocity field $\mathbf{v} = (a, b)$ is constant. In other words, both a and b are constant two-variable functions in the (x, y) plane.

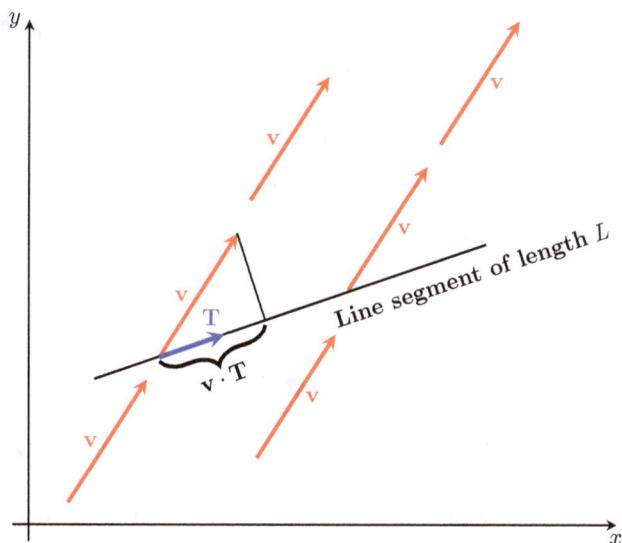

Figure 3.3   Component of $\mathbf{v}$ along a line segment of length L.

Suppose that a line segment of length L has a unit tangent vector $\mathbf{T} = (c, d)$, (that is $c^2 + d^2 = 1$).

Note: The fact that components of velocity vectors are relevant is a physical result. That is, in applications, velocity and momenta vectors add component-wise.

And so, making use of the dot product, we can compute the mass flow rate along the length of L. For then the component of $\mathbf{v} = (a, b)$ in the direction of L (parallel to L) is $\mathbf{v} \cdot \mathbf{T}$. Hence, the mass-flow rate along all of L is

$$(\mathbf{v} \cdot \mathbf{T}) \text{ (length of line segment)}.$$

3.3   Circulation along a curve in 2-d

For simplicity, we restrict attention to curves C that are one-one and differentiable. (See a Calculus III textbook for examples.)

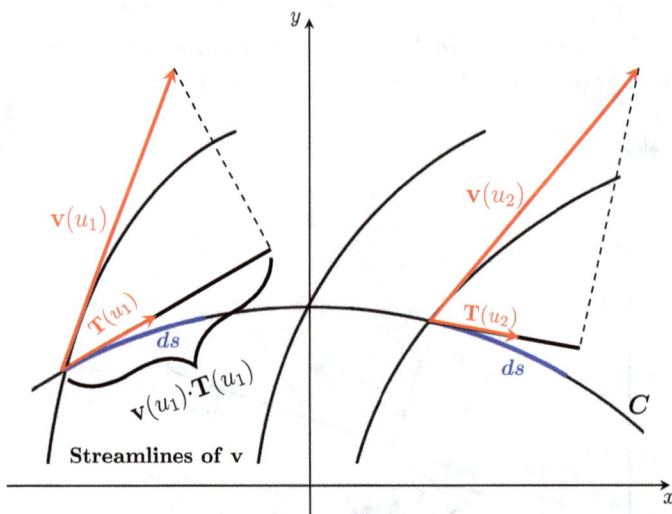

Figure 3.4   Element of circulation along a curve C is $d(circulation)|_C = \mu((\mathbf{v} \cdot \mathbf{T})ds)$.

Comment: As you might expect, the key is repeated use of linear approximation.

Suppose that a velocity field $\mathbf{v} = \Big(a(x, y), b(x, y)\Big)$ and a curve C is given by coordinate functions $\mathbf{r}(u) = \Big(x(u), y(u)\Big)$.

The problem then is to determine the total rate of mass flow parallel to the curve C and along its entire length.

Prior to dealing with details of parameterization, let's begin with some basic geometry.

An element of arc-length is ds and $\mathbf{T}$ is the unit tangent vector along the curve.

The corresponding *element of circulation* (i.e., to first-order approximation, the total mass flow rate in the direction of curve element of length ds) is

$$d(circulation)|_C = \mu((\mathbf{v} \cdot \mathbf{T})ds).$$

This leads to the following definition:

Definition 3.1. *Total circulation along* C is the limit of sums of elements of circulation. That is, the *circulation* of $\mathbf{v}$ along C is

$$\int_C (\mathbf{v} \cdot \mathbf{T})\, ds := \lim_{ds \to 0} \sum_{i=1}^{k} (\mathbf{v} \cdot \mathbf{T})\, ds.$$

By reviewing "curves" in a Calculus III textbook, this integral is computed by using a parameterization for the curve C. See Figure 3.4, above. In some cases, that parameter can be arc-length s. However, we need a symbol other than s to allow for all cases. It also would be good to have a symbol different from t. That is because, in fluid dynamics, t is used to represent 'time.' However, in the present context, the curve C is a fixed curve. It is the fluid, not C, that is in motion.

To avoid ambiguity, we use a different but also common symbol for a parameter for curves, namely, u.

Suppose that we have a parameterization for the curve C. In order to compute the integral in Definition 3.1, let's first sort out the integrand.

As you will remember from Calculus III, using first-order approximation we get

$$ds = \| \mathbf{r}'(u) \| \, du = \| (x'(u), y'(u)) \| \, du,$$

where the derivative is with respect to the parameter u.

A unit tangent vector to the curve is

$$\mathbf{T} = \frac{(x'(u), y'(u))}{\| (x'(u), y'(u)) \|}.$$

Along the curve

$$\mathbf{v}(u) = \Big(a(x(u), y(u)), b(x(u), y(u)) \Big), \quad \text{where } u_0 \leq u \leq u_1.$$

Substituting these into the defining integral, we get that the *total circulation of* **v** *along the curve* C is

$$\int_{u_0}^{u_1} \left[\frac{\Big(a(x(u), y(u)), b(x(u), y(u))\Big) \cdot (x'(u), y'(u))}{\| (x'(u), y'(u)) \|} \right] \| (x'(u), y'(u)) \| \, du$$

$$= \int_{u_0}^{u_1} \left[\Big(a(x(u), y(u)), b(x(u), y(u))\Big) \cdot (x'(u), y'(u)) \right] du$$

Exercises

In Exercises 3.3.1–3.3.13, use the formula in Section 3.3 to compute the total circulation for each velocity $\mathbf{v}(x, y)$ along the curve C. Graph and explain your results.

Exercise 3.3.1
$\mathbf{v}(x, y) = (3, 0)$, C is the line-segment along the x-axis that starts at $(0, 0)$ and ends at $(5, 0)$.

Exercise 3.3.2
$\mathbf{v}(x, y) = (3, 0)$, C is the line-segment along the x-axis that starts at $(0, 0)$ and ends at $(0, 5)$.

Exercise 3.3.3
$\mathbf{v}(x, y) = (x, 0)$, C is the line-segment along the x-axis that starts at $(0, 0)$ and ends at $(5, 0)$.

Exercise 3.3.4
$\mathbf{v}(x, y) = (x, e^{x+y})$, C is the line-segment along the x-axis that starts at $(0, 0)$ and ends at $(5, 0)$.

Exercise 3.3.5
$\mathbf{v}(x, y) = (x, b(x, y))$, where $b(x, y)$ is any function for which the integrals exist (e.g., continuous), and C is the line-segment along the x-axis that starts at $(0, 0)$ and ends at $(5, 0)$.

Exercise 3.3.6

$\mathbf{v}(x, y) = (3, 0)$, C is the boundary of the square with vertices $A(0, 0)$, $B(1, 0)$, $C(1, 1)$, $D(0, 1)$, parameterized counter-clockwise and starts at $A(0, 0)$.

Exercise 3.3.7

$\mathbf{v}(x, y) = (x, 0)$, C is the boundary of the square with vertices $A(0, 0)$, $B(1, 0)$, $C(1, 1)$, $D(0, 1)$, parameterized counter-clockwise and starts at $A(0, 0)$.

Exercise 3.3.8

$\mathbf{v}(x, y) = (x, y)$, C is the unit circle centered at the origin $(0, 0)$, initial point $A(1, 0)$, parameterized counter-clockwise.

Exercise 3.3.9

$\mathbf{v}(x, y) = (x, x)$, C is the unit circle centered at the origin $(0, 0)$, initial point $A(1, 0)$, parameterized clockwise.

Exercise 3.3.10

$\mathbf{v}(x, y) = (-y, x)$, C is the unit circle centered at the origin $(0, 0)$, initial point $A(1, 0)$, parameterized counter-clockwise.

Exercise 3.3.11

$\mathbf{v}(x, y) = (-y, x)$, C is the circle of radius $r > 0$ centered at the origin $(0, 0)$, initial point $A(1, 0)$, parameterized counter-clockwise.

Exercise 3.3.12

$\mathbf{v}(x, y) = (3x, x)$, C is the boundary of the triangle with vertices $P(0, 0)$, $Q(10/3, 0)$, $R(3, 0)$, initial point $P(0, 0)$, parameterized counter-clockwise.

Exercise 3.3.13

$\mathbf{v}(x, y) = (3x + y, x + y)$, C is the boundary of the triangle with vertices $P(0, 0)$, $Q(10/3, 0)$, $R(3, 0)$, initial point $P(0, 0)$, parameterized counter-clockwise.

3.4 Invariance of the circulation integral

Evaluation of this integral depends on having a parameterization for the curve C. You might wonder, what happens when we use different parametrizations?

To explore that question, let's first look to a simpler case, where the curve is along the x-axis and $a(x(u), y(u)) = 0$. In that special case, the integral is of the form

$$\int_{u_0}^{u_1} b(y(u))y'(u)du.$$

Does this remind you of something from an earlier mathematics course?

The parameterization of the curve is the "inside" function $y(u)$. From Calculus I, think of the chain rule, and its applications to integration called the "method of substitution." That method works because

$$b(y(u))y'(u) = \frac{d}{du}b(y(u)).$$

In other words, the chain rule is what allows us to evaluate an integral using a parameter that happens to be convenient. So long as the criteria of the chain rule are met, the integral has the same value, namely,

$$B(y(u_1)) - B(y(u_0)), \text{ where } \frac{d}{dy}B(y) = b(y)$$

and $y(u_0)$ and $y(u_1)$ are, by definition, the same for all parameterizations, because these represent the y-coordinates of the end points of the curve C in the ambient coordinate plane.

Problem 3.2.
A good exercise is sort this out in the general case, without appealing to a textbook. It is a good way to shore up one's understanding of the chain rule. In other words, show, by computation, that the circulation integral does not depend on parameterization.

3.5   Abbreviated notation for the circulation integral

Thanks to the invariance just discussed, we obtain the following (unambiguous) abbreviation for *circulation*:

Circulation of $\mathbf{v}$ *along* C is

$$\int_C adx + bdy = \int_C (a, b) \cdot (dx, dy).$$

Another convenient abbreviation is obtained by recalling that

$$\mathbf{v}(u) = \Big(a(x(u), y(u)), b(x(u), y(u))\Big),$$

and that (dx, dy) is an increment in the plane that can be written as $d\mathbf{r}$. That is, $\mathbf{r} := (x, y)$ and $d\mathbf{r} := (dx, dy)$.

Using this more concise notation, we can also write *circulation of* $\mathbf{v}$ along C is

$$\int_C \mathbf{v} \cdot d\mathbf{r}.$$

In order to compute circulation, we need a parameterization of the curve. But in these last two formulas, the parameterization is nowhere in sight. Why is that? At this stage, the abbreviation means: introduce whatever parameterization is convenient. This is valid because as in Problem 3.2, circulation is (within a class of parameterizations) independent of parameterization.

With parameter u and Leibniz's notation, dx becomes

$$\frac{dx}{du} du$$

and dy becomes

$$\frac{dy}{du} du.$$

The other terms are $a(x(u), y(u))$ and $b(x(u), y(u))$.

With these substitutions, one then computes a real integral, just as in Calculus I.

Observation: In a more advanced course, the formula for circulation is a "function" of two "variables," namely, C and $\mathbf{v}$. Why is that? It is because circulation is independent of parameterization. Clearly, however, it depends on the curve C and the vector field $\mathbf{v}$, both defined on the plane.[27]

3.6   Adding circulations from different curves

Suppose that, in order to collect water for towns A and B, a hydro-engineering firm plans to install a pipe system across a river (see Figure 3.5).

---

[27]In a more advanced context ("higher viewpoint" — see Part A of the Supplement), terms of the form $w = a\,dx + b\,dy$ are called "1-forms" and C is a "one-dimensional manifold." Higher dimensional objects like surfaces, volumes, and so on are "two-dimensional manifolds," "three-dimensional manifolds," and so on. The "k-forms" formalize the notion of "line element," "area element" and "volume element," respectively. A one-form $w = a\,dx + b\,dy$ provides "line elements" along a curve C in the sense that, at each point p on C, it defines a linear functional on the space of tangent vectors at p. Computation reduces to what we have described above. That is, one needs to introduce a parameterization and integrate in the usual way.

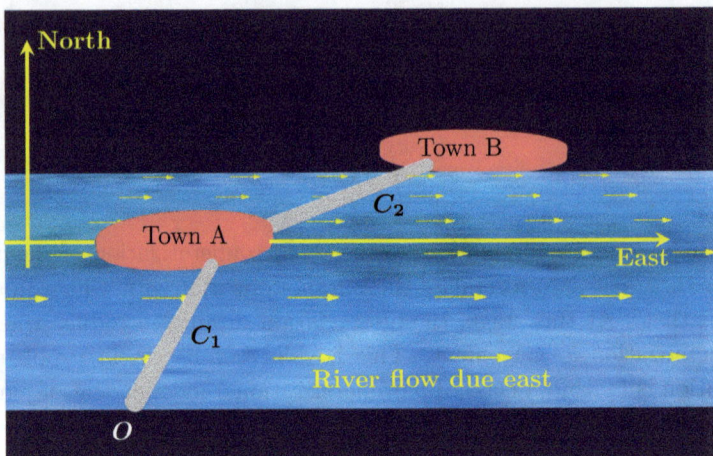

Figure 3.5   Two pipes collecting river water that flows due east.

Let's suppose that the river flows west to east. Town A is on an island in the middle of the river and town B is on the north shore, somewhat east, and so downstream, of the island. A first pipe will connect a location O on the southern shoreline to town A. That will provide the island-town with water. A second pipe will connect the island to the northern shoreline. That pipe will provide water for town B. There will be collection holes in both pipes into which river water flows. We are ignoring numerous aspects of the engineering problem. Still, the rate at which water flows into and then along a collection pipe, while not identical to, is proportional to a circulation integral.

The part of the problem that is new here is that we will need to compute *two* integrals. Let $\mathbf{v}$ be the vector field for the river velocity, C_1 is the curve across the river that the first pipe will follow from the southern shore to the island; and C_2 is the curve across the river that the second pipe will follow, connecting the island to the northern shore.

At a moment in time, we consider the sum of all mass-flow rates at all locations along both pipes. So, the total mass-flow rate of water along the two pipes will be the sum of two circulation integrals. However, some care is needed in choosing parameterizations.

Suppose that, for the pipe connecting the southern shore to the island, the circulation integral is parameterized so that the initial point is on the southern shore and the final point is on the island.

Then, one contribution to total circulation is

$$\int_{C_1} adx + bdy = \int_{C_1} (a, b) \cdot (dx, dy) = \int_{C_1} \mathbf{v} \cdot (dx, dy).$$

The circulation integral for the second pipe is

$$\int_{C_2} adx + bdy = \int_{C_2} (a, b) \cdot (dx, dy) = \int_{C_2} \mathbf{v} \cdot (dx, dy).$$

We already know that circulation is independent of parameterization. Might that also work for *two* line integrals?

Note, however, that invariance of the circulation integral has hypotheses. Parameterizations are required to have the same initial end points. That invariance does not tell us how to combine two independently defined circulations along different curves with initial and end points determined by an application.

Clues can be had by looking to familiar integrals. In Calculus II, an integral over an interval $[c, d]$ is an integral over a curve, namely, a parameterized line segment where the initial point is c and the final point is d.

What happens to the integral if we reverse that, so that the initial point of the integral is d and the final point is c. In other words, how are the following two integrals related,

$$\int_c^d f(x)dx \text{ and } \int_d^c f(x)dx?$$

You may remember that they are "the same" but opposite in sign.[28]

If you trace back through how to compute circulation, you will find the same result. That is, if we switch end points, the sign is reversed, in the original circulation. To see that, introduce a parameterization and you will have a Calculus II problem.

What does that tell us about the engineering problem for the two towns?

If the engineering firm uses the island as the end point of both C_1 and C_2, then one of the circulations will be negative in sign to the other. We leave it as an exercise for the reader to explore why that would cause problems. One obvious problem is that adding two circulations where the connecting curves

---

[28] *Hint*: An original notion of integral is "area under a curve, over an interval." When intervals change, how do areas add or subtract?

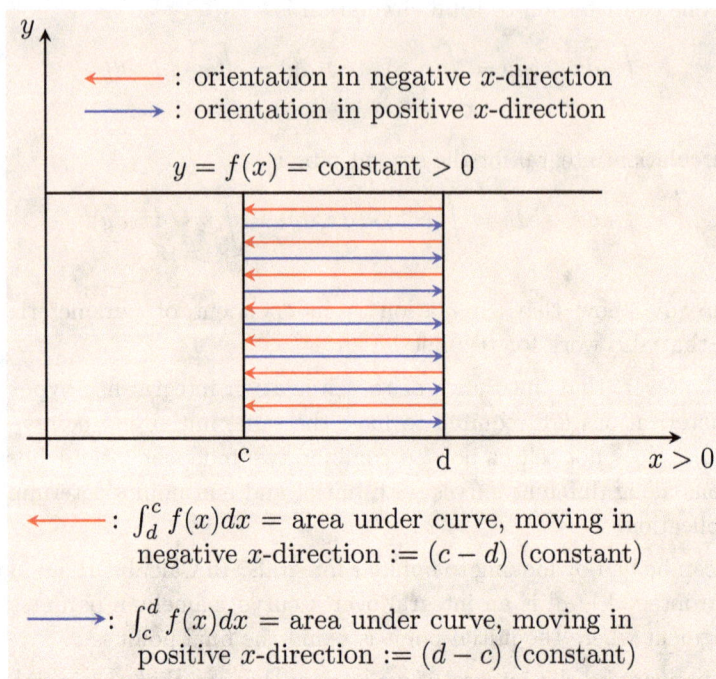

Figure 3.6   Elementary integration. Reversing order of integration changes sign of integral.

are "oppositely oriented" can result in a computed zero for circulation. Of course, that would not match the situation of an actual river flow along two pipes.

How, then, should we combine the two circulation integrals, the one for C_1 and the other for C_2. We already have the invariance result. We can choose parameterizations for each curve, individually.

But we need to combine them in a way that we avoid the problem just described and so that the sum provides quantities needed in hydro-engineering.

There is only one degree of freedom.

That is, we need the end of point of C_1 to be the initial point of C_2.

Doing that, we obtain the following definition for adding curves and circulations.

Definition 3.2. Suppose that the end point of a curve C_1 is the initial point of curve C_2. Then, the sum of the two curves is denoted by $C = C_1+C_2$. By definition, the "sum" means: join the two curves, as described above, to get a new "larger" curve. The circulation along the sum of curves is then defined to be

$$\int_{C_1+C_2} \mathbf{v} \cdot d\mathbf{r} = \int_{C_1} \mathbf{v} \cdot d\mathbf{r} + \int_{C_2} \mathbf{v} \cdot d\mathbf{r}.$$

Note: This result extends to adding any finite number of curves. It follows that

$$\int_{C_1+C_2+\cdots+C_m} \mathbf{v} \cdot d\mathbf{r} = \int_{C_1} \mathbf{v} \cdot d\mathbf{r} + \int_{C_2} \mathbf{v} \cdot d\mathbf{r} + \cdots + \int_{C_m} \mathbf{v} \cdot d\mathbf{r},$$

so long as the end point of C_i is the initial point of C_{i+1}, for $i = 1, \cdots, m-1$.

Problem 3.3.
Prove that under the hypotheses, the definition for adding circulation integrals is independent of parameterizations.

Exercises

Exercise 3.6.1
Let $\mathbf{v}(x,y) = (x,0)$ and C be the boundary of the square with vertices $P(0,0)$, $Q(1,0)$, $R(1,1)$, $S(0,1)$, parameterized counter-clockwise and starts at $P(0,0)$. Using the four sides of the square as curves, write C as a sum $C = C_1 + C_2 + C_3 + C_4$. Show that

$$\int_C \mathbf{v} \cdot d\mathbf{r} = \int_{C_1} \mathbf{v} \cdot d\mathbf{r} + \int_{C_2} \mathbf{v} \cdot d\mathbf{r} + \int_{C_3} \mathbf{v} \cdot d\mathbf{r} + \int_{C_4} \mathbf{v} \cdot d\mathbf{r}.$$

Exercise 3.6.2
Let C_1 be the unit circle centered at the origin $(0,0)$, initial point $A(1,0)$, parameterized counter-clockwise, and let C_2 be the unit circle centered at the origin $(0,0)$, initial point $A(1,0)$, parameterized clockwise.

(a) Describe $C_1 + C_2$.

(b) Let $\mathbf{v}(x, y) = (a(x, y), b(x, y))$ be a velocity field. Show that

$$\int_{C_1} \mathbf{v} \cdot d\mathbf{r} + \int_{C_2} \mathbf{v} \cdot d\mathbf{r} = \int_{C_1 + C_2} \mathbf{v} \cdot d\mathbf{r}.$$

3.7   Preparing for Green's Theorem

Our inquiry began with preliminary descriptions of fluid, such as water, in motion. An example is a "steady-state," that is, when water flows with constant velocity in a shallow chute or canal. Other examples include: flow of a shallow river where water moves faster in the center of the river while near the shoreline water speed slows to almost zero; the spiral motion of water in a whirlpool; and the outward flow of water from a source.

In all of these cases, surfaces can be seen to rotate, contract and dilate, as the case may be.

Simplified versions of such patterns are captured in the following mathematical examples.

In each case, we compute an approximation to circulation about a small rectangle of dimensions $dx \times dy$.

By appealing to elementary calculus, we will get a general result. That result will provide us with the ingredients needed for Green's Theorem. For the moment, however, let's focus on the following examples.

Example
In this example, let $\mathbf{v} = (2, 0)$.

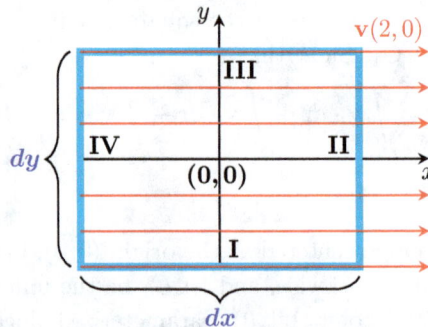

Figure 3.7   Vector field $\mathbf{v} = (2, 0)$ and rectangle $dx \times dy$ centered at $(0, 0)$ with sided **I**, **II**, **III** and **IV**.

Circulation is a mass-flow rate along a curve. However, let's explore the idea of "circulation about a point" which, in this case, is $(0,0)$.

Sides **I** and **III** are parallel to the x-axis, while sides **II** and **IV** are parallel to the y-axis. The net mass flow rate is, then, a combination of circulations along **I** and **III**, respectively. However, what combination should that be?

We need to compute along curves that are appropriately connected. Start with **I** parameterized in the positive x-direction. Tracking around the rectangle to compute circulation along all segments that surround the origin, we get

[Circulation along **I**] + [Circulation along **II**] + [Circulation along **III**]

+ [Circulation along **IV**]

with the stipulation that these are parameterized so that the end point of **I** is the initial point of **II**, the end point of **II** is the initial point of **III** and the end point of **III** is the initial point of **IV**.

Problem 3.4.
Show that this yields the following sum:

$$[2dx] + [0] + [-2dx] + [0] = 0.$$

Why are there zeros? It is because the vector field is orthogonal to increments **II** and **IV**, respectively. This is consistent with the graphed vector field. There is no circulation *about* the origin because the flow is constant in one direction. This is only descriptive, but it is a clue.

Example
In this example, let $\mathbf{v} = (y, 0)$.

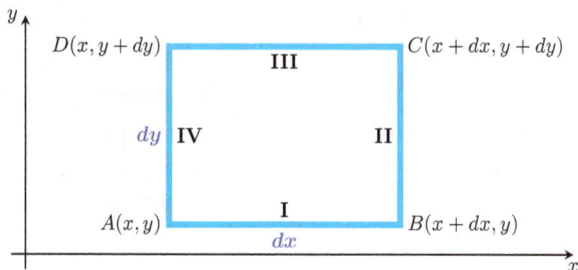

Figure 3.8 Vector field $\mathbf{v} = (y, 0)$. Rectangle in the first quadrant with vertices $A(x, y)$, $B(x + dx, y)$, $C(x + dx, y + dy)$ and $D(x, y + dy)$.

Draw a rectangle in the first quadrant of the (x, y) plane with the following vertices: $A(x, y)$, $B(x + dx, y)$, $C(x + dx, y + dy)$ and $D(x, y + dy)$.

Appropriately parameterized, AB is **I**, BC is **II**, DC is **III** and DA is **IV**.

Let's now compute circulation along the boundary of the rectangle, using the curve **I + II + III + IV**, as in the diagram.

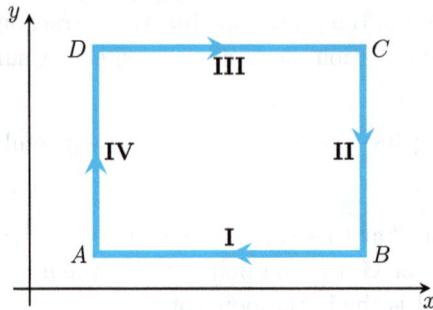

Figure 3.9   Boundary of rectangle parameterized: **I + II + III + IV**, where **I** = AB, **II** = BC, **III** = CD and **IV** = DA.

The computation gives

$$[ydx] + [0] + [(y + dy)(-dx)] + [0] = -dydx.$$

What does the negative sign mean?

In the diagram, the vertices of **IV** are A and D.

The vector field $\mathbf{v} = (y, 0)$ has only horizontal components.

For a small time increment, $dt > 0$ say, A and D move to A' and D', respectively.

These are determined by following "test points" along streamlines. So, AD and $A'D'$ are along "lines of sight" obtained when following A and D to A' and D', respectively. And, we can compare line segments AD and $A'D'$.

Remark regarding rotation: In the last example, relative to the y-axis, the angle of the line of sight between the two vertices changes. More precisely, a negative (i.e., clockwise) angle $d\theta$ is introduced.

Is it a coincidence that *circulation* and $d\theta$ are both negative? [29]

For now, let's leave the observation as needing further inquiry.

---

[29] One way to sort out what the relationship is to invoke Green's Theorem. But that would be making use of a result that in this book has not yet been established.

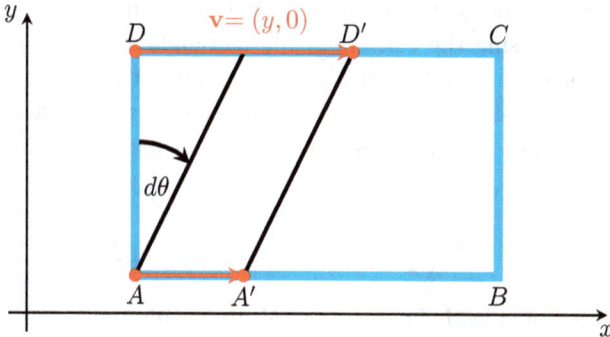

Figure 3.10 Small time increment $dt > 0$, A flows to A', and D flows to D'. The vector field is $\mathbf{v} = (y, 0)$ implies $|DD'| > |AA'|$.

Example

For this example, $\mathbf{v} = (-y, x)$ is the familiar vector field whose streamlines orbit the origin $(0, 0)$.

As in previous examples, we can introduce a small rectangle which, in this example, will be centered about $(0, 0)$.

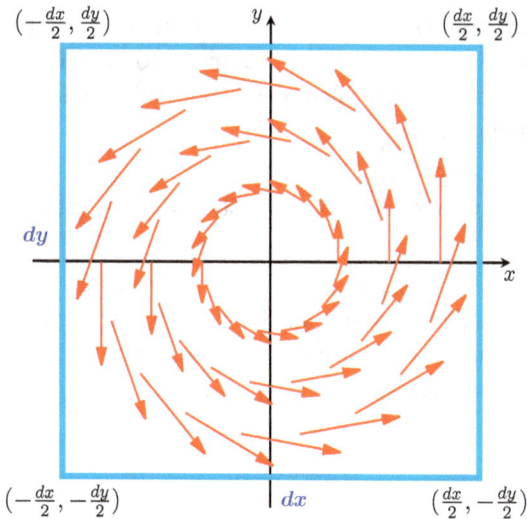

Figure 3.11 Vector field $\mathbf{v} = (-y, x)$. Small rectangle $dx \times dy$ centered at $(0, 0)$.

The rectangle of dimensions $dx \times dy$ does not easily fit into circular stream-lines. Instead of using the corners of the rectangle as reference points, as in the previous examples, let's use the mid-points of each side of the rect-angle. This is equally valid since all of our computations are by way of approximation. The lengths of each side will be dx and dy, respectively.

The vertices of the rectangle need to be:

$$A\left(-\frac{dx}{2}, -\frac{dy}{2}\right), B\left(\frac{dx}{2}, -\frac{dy}{2}\right), C\left(\frac{dx}{2}, \frac{dy}{2}\right) \text{ and } D\left(-\frac{dx}{2}, \frac{dy}{2}\right).$$

The mid-points of sides **I**, **II**, **III** and **IV** are then

$$P\left(0, -\frac{dy}{2}\right), Q\left(\frac{dx}{2}, 0\right), R\left(0, \frac{dy}{2}\right) \text{ and } S\left(-\frac{dx}{2}, 0\right).$$

The vector field is defined by $\mathbf{v}(x, y) = (-y, x)$. Hence,

$$\mathbf{v}(P) = \left(\frac{dy}{2}, 0\right), \mathbf{v}(Q) = \left(0, \frac{dx}{2}\right), \mathbf{v}(R) = \left(-\frac{dy}{2}, 0\right) \text{ and } \mathbf{v}(S) = \left(0, -\frac{dx}{2}\right).$$

Therefore, the approximation to circulation around the rectangle ABCD is

$$\left(\frac{dy}{2}\right) dx + \left(\frac{dx}{2}\right) dy + \left(-\frac{dy}{2}\right)(-dx) + \left(-\frac{dx}{2}\right)(-dy) = 2dxdy.$$

Remark regarding rotation: As in the previous example, for a small time increment, $dt > 0$ say, A and B move to A' and B'. Notice that once again, for a small time increment, algebraic signs of circulation and an angular change are the same.

Example

For this example, we model a source with $\mathbf{v} = (x, y)$. Streamlines emanate from the origin.

Again, construct a rectangle $ABCD$, with dimensions as in the previous example, centered about the origin. Show that, in this case, the circulation is zero. Can you anticipate from the diagram why this is so?

Now that we have explored a few examples, let's work on the general case.

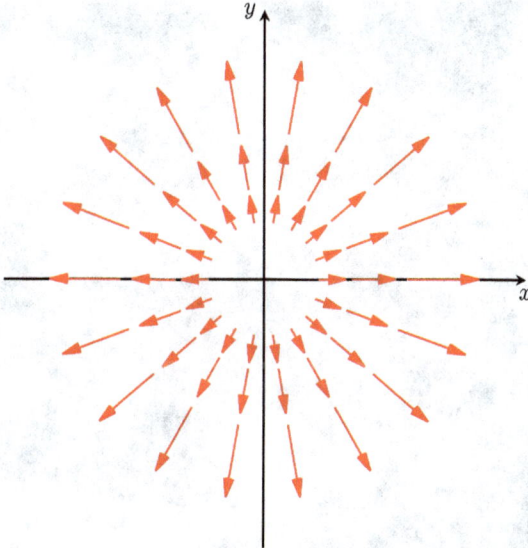

Figure 3.12   Vector field $\mathbf{v} = (x, y)$.

Circulation about a rectangle with dimensions dx × dy and consequences of linear approximation

The vector field is $\mathbf{v}\Big(a(x, y), b(x, y)\Big)$. The vertices of the rectangle $ABCD$ are

$$A(x, y), \ B(x + dx, y), \ C(x + dx, y + dy) \ \text{and} \ D(x, y + dy).$$

To approximate circulation around the rectangle, we compute:

$$a(x, y)dx + b(x + dx, y)dy + a(x + dx, y + dy)(-dx) + b(x, y + dy)(-dy).$$

Grouping terms, we obtain

$$[b(x + dx, y) - b(x, y + dy)]dy - [a(x + dx, y + dy) - a(x, y)]dx.$$

The next step is elementary and is the key to discovering Green's Theorem. It leads to the main ingredient for the theorem. See Section 3.8.

Grouping terms, there are differences in function values. And linear approximation tells us how to approximate those differences.

In two variables, linear approximation involves two partial derivatives and a reference point. In the present application, the reference point is (x, y).

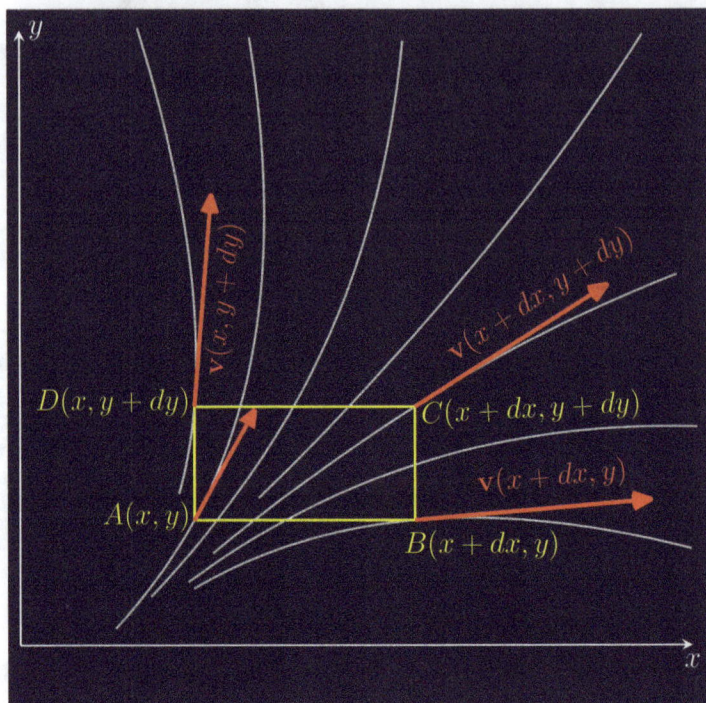

Figure 3.13 $\mathbf{v}(x, y) = (a(x, y), b(x, y))$, evaluated at vertices $A(x, y)$, $B(x+dx, y)$, $C(x+dx, y+dy)$ and $D(x, y+dy)$.

But in the two differences

$$[b(x + dx, y) - b(x, y + dy)] \quad \text{and} \quad [a(x + dx, y + dy) - a(x, y)],$$

only one of the terms involves the reference point (x, y). A common trick (which survives into higher analysis) is to compare by getting from one location to the next in steps that, strategically, "go through" the reference point (x, y). For instance, one can do this by following the edges of the rectangle.

In the difference

$$b(x + dx, y) - b(x, y + dy),$$

the "final point" is $(x + dx, y)$ while the "initial point" is $(x, y + dy)$. To compare these points, first compare the function values at $(x + dx, y)$ and (x, y); and then compare the function values at (x, y) and $(x, y + dy)$.

This means that we compute

$$b(x + dx, y) - b(x, y + dy) = b(x + dx, y) - b(x, y) + b(x, y) - b(x, y + dy).$$

Grouping terms we get,

$$b(x + dx, y) - b(x, y + dy) = \big(b(x + dx, y) - b(x, y)\big) + \big(b(x, y) - b(x, y + dy)\big).$$

This gives

$$\frac{\partial b}{\partial x}(x, y)dx + \frac{\partial b}{\partial x}(y, y + dy)(-dy).$$

Remember that this is an approximation only for the first term in brackets in the expression

$$[b(x + dx, y) - b(x, y + dy)]dy - [a(x + dx, y + dy) - a(x, y)]dx.$$

We will need to sort out the second bracket too. But for the moment, let's continue with our focus on the first square bracket, and substitute what we just computed.

The first square bracket becomes

$$\left[\frac{\partial b}{\partial x}(x, y)dx + \frac{\partial b}{\partial x}(y, y + dy)(-dy) \right] dy$$

$$= \frac{\partial b}{\partial x}(x, y)dxdy + \frac{\partial b}{\partial x}(y, y + dy)(-dy)dy.$$

In the present context, then, circulation is found to be an integral relative to an area of a small rectangle, relative to which the term

$$\frac{\partial b}{\partial y}(y, y + dy)dydy \approx 0.$$

(This needs to be made precise. It can be shown that the term

$$\frac{\partial b}{\partial y}(y, y + dy)dydy$$

does not contribute to the integral. In a more advanced course, we would need a limit argument.)

What do we have so far?

With linear approximation,

$$[b(x + dx, y) - b(x, y + dy)]dy \approx \frac{\partial b}{\partial x}(x, y)dxdy.$$

Problem 3.5.

Work out the details and show that, in our linear approximation, the other square bracket gives a similar result, namely:

$$-[a(x+dx, y+dy) - a(x,y)]dx \approx -\frac{\partial a}{\partial y}(x,y)dxdy.$$

If we combine the two sets of computations, what is the result? (See (2) below.)

We now have two ways to approximate circulation of $v(x,y)$ along the positively oriented boundary of a small rectangle $dx \times dy$

(1) A direct approximation is

$$[b(x+dx, y) - b(x, y+dy)]dy - [a(x+dx, y+dy) - a(x,y)]dx.$$

(2) If we use linear approximation for differences in function values in 1 (ignoring higher order terms involving $dxdx$, $dyddy$, $dxdxdy$, $dxdydy$, and so on, coming from Taylor's theorem) we obtain a different approximation to circulation along the boundary of the small rectangle, namely,

$$\left[\frac{\partial b}{\partial x}(x,y) - \frac{\partial a}{\partial y}(x,y)\right]dxdy.$$

Higher order terms do not enter into the final integrals.

Written as a single equation (ignoring higher order terms), we have two ways of computing circulation around the small rectangle $dx \times dy$. That is,

$$[b(x+dx, y) - b(x, y+dy)]dy - [a(x+dx, y+dy) - a(x,y)]dx$$
$$= \left[\frac{\partial b}{\partial x}(x,y) - \frac{\partial a}{\partial y}(x,y)\right]dxdy$$

This equality is the key to Green's Theorem.

Before getting to Green's theorem, however, an observation is already possible.

Circulation around the small rectangle $dx \times dy$ is proportional to the area of the small rectangle, namely, the product $dxdy$, where the proportionality factor is

$$\frac{\partial b}{\partial x}(x,y) - \frac{\partial a}{\partial y}(x,y).$$

In other words, dividing by $dxdy$, the term $b_x - a_y$ is *circulation density per unit area* (with the above construction implicit).

3.8 Green's Theorem

From Section 3.7, using linear approximation, circulation around a small rectangle is equal to the area of the rectangle times the *circulation density function*, namely, $b_x - a_y$.

In practice, we need to know circulation around not merely a small rectangle, but a region R whose *positively oriented* boundary ∂R is a closed curve C.

Recall that a curve C is closed when its initial point is the same as its final point. To rigorously define positive orientation for a curve would go beyond the context of this introduction. For present purposes, a parameterization is positive when it tracks counter-clockwise relative to an interior point of the region R.

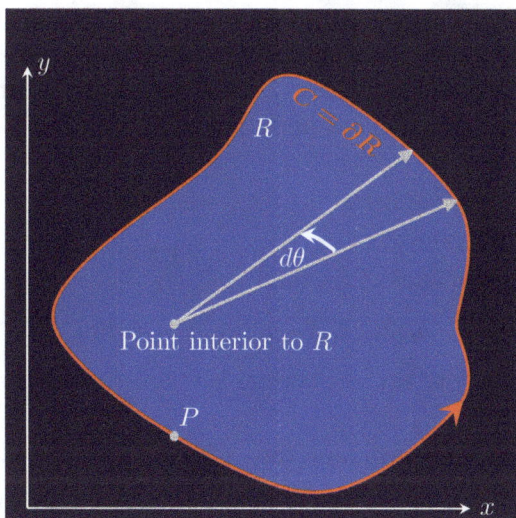

Figure 3.14 Closed curve, $P = \mathbf{r}(u_0) = \mathbf{r}(u_1)$, with positive orientation, that is, parameterization tracks boundary curve $C = \partial R$ counter-clockwise, with $d\theta > 0$.

Question: Can our result about a small rectangle be extended to a positively oriented boundary curve C of a region R?

We can take advantage of our result from the previous subsection:

$$[b(x + dx, y) - b(x, y + dy)]dy - [a(x + dx, y + dy) - a(x, y)]dx$$

$$= \left[\frac{\partial b}{\partial x}(x, y) - \frac{\partial a}{\partial y}(x, y)\right] dxdy$$

In other words,

[circulation of $\mathbf{v} = (a, b)$ around the boundary of a rectangle $dx \times dy$]

$$= [b_x - a_y] \, dxdy.$$

The term

$$[b_x - a_y] \, dxdy$$

can be extended to the region R. This is because it is a two variable function multiplied by an area element $dxdy$. In other words, it is an integrand for an integral over the region R.

This is our clue.

By definition of an integral in the plane, we introduce a grid of small rectangles that partition the region R. We restrict to that part of the partition that lies entirely within the region R. We can add up the results and then take the limit, where the limit is as dimensions of the small rectangles converge to zero. All of this is made precise in Calculus III and in more advanced courses.

To explore this in more detail, let's see what happens if we look at four sub-rectangles in a grid.

There are four summands of the form

$$[b_x(x_{ij}, y_{ij}) - a_y(x_{ij}, y_{ij})]dxdy,$$

where each (x_{ij}, y_{ij}) is, say, the bottom left corner of each rectangle in the sum.

Where does circulation come into play?

Remember that in their original form (before linear approximation), each of those summands is an approximation to

[circulation of $\mathbf{v} = (a, b)$ around the boundary of a rectangle $dx \times dy$].

And, if we add up area elements, then we get a corresponding sum of circulations.

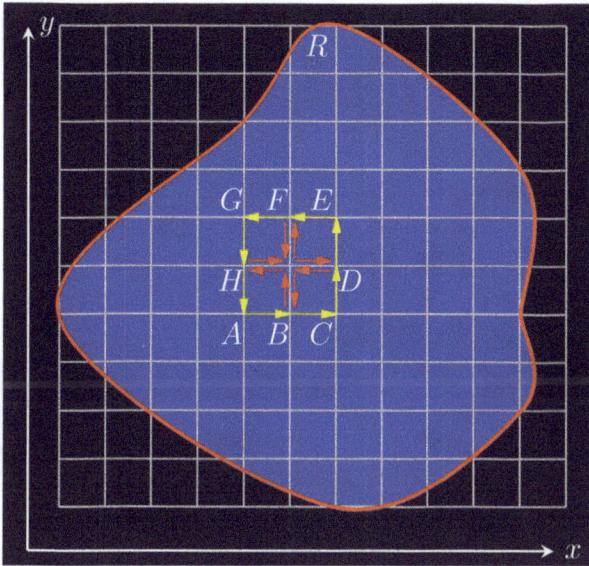

Figure 3.15   Four sub-rectangles of a partition of R.

Problem 3.6.

As the diagram suggests, contributions to circulation along common edges internal to the four rectangles all cancel. That is because they are each computed twice, with opposite signs.

Hence .... !

Do you see it?

The sum of circulations around the four rectangles reduces to the circulation along the outer edge of the large rectangle that is the union of the four small rectangles.

Hence .... !!

And, we can keep going!!!

That is, if we add up circulations around all rectangles in the partition within the region R, common internal terms cancel. And this leaves circulation along the closed curve that is determined by the union of the outer edges of rectangles of the partition that lie within R.

In the limit, these outer edges converge to the boundary curve of the region and the sums therefore converge to the contour integral along the closed

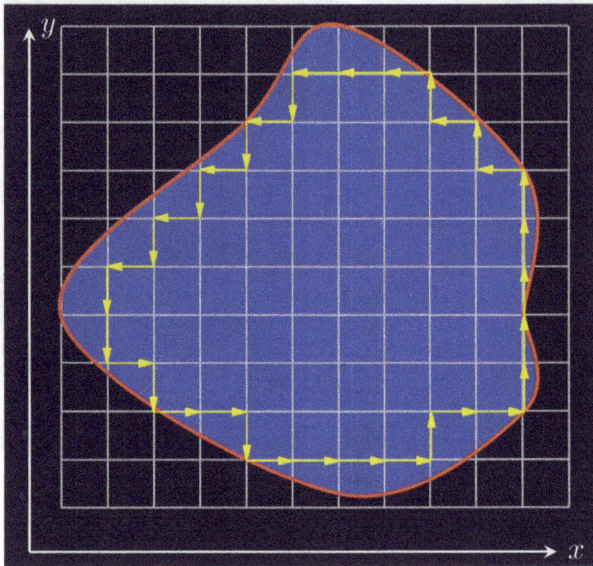

Figure 3.16   (See Figure 3.15.) Interior circulations cancel. Remaining terms are along boundary of grid that is interior to the region R.

curve,

$$\oint_C \mathbf{v} \cdot dr.^{30}$$

On the other hand, linear approximation of each summand means that we also end up with sums that converge to an area integral, namely,

$$\iint_R [b_x - a_y] dx dy.$$

The two computations are two different ways of computing the same quantity.

And so we get the following:

$$\oint_C \mathbf{v} \cdot d\mathbf{r} = \oint_C a dx + b dy = \iint_R [b_x - a_y] dx dy$$

This equality is called Green's Theorem.

Below, we go on to a different question that also yields the quantity $b_x - a_y$.

----

[30]Steps need proof including rigorous limit arguments. See Preface.

In that context we will find out that "curl" is a fitting name for the quantity $b_x - a_y$ which, as above, is also seen to be the *circulation density*.

This means that Green's Theorem can also be written

$$\oint_{\partial R} \mathbf{v} \cdot d\mathbf{r} = \iint_R (\operatorname{curl} \mathbf{v}) dx dy$$

where ∂R stands for the positively oriented boundary of R, which in the formula above is also called C and, as is customary, the "circle" on the integral symbol indicates that C is closed.

Examples

(1) For $\mathbf{v} = (2, 0)$, circulation density of $\mathbf{v} = 0 + 0 = 0$.

(2) For $\mathbf{v} = (y, 0)$, circulation density of $\mathbf{v} = -1$.

(3) For $\mathbf{v} = (-y, x)$, circulation density of $\mathbf{v} = 1 - (-1) = 2$.

(4) For $\mathbf{v} = (x, y)$, circulation density of $\mathbf{v} = 1 - 1 = 0$.

Exercises

In Exercises 3.8.1–3.8.11, compute both sides of Green's Theorem, for each velocity $\mathbf{v}(x, y)$ along the curve $C = \partial R$ that is the boundary of the region R.

Exercise 3.8.1

$\mathbf{v}(x, y) = (3, 0)$, C is the boundary of the square with vertices $A(0, 0)$, $B(1, 0)$, $C(1, 1)$, $D(0, 1)$, parameterized counter-clockwise and starts at $A(0, 0)$.

Exercise 3.8.2

$\mathbf{v}(x, y) = (x, 0)$, C is the boundary of the square with vertices $A(0, 0)$, $B(1, 0)$, $C(1, 1)$, $D(0, 1)$, parameterized counter-clockwise and starts at $A(0, 0)$.

Exercise 3.8.3

$\mathbf{v}(x, y) = (x, y)$, C is the unit circle centered at the origin $(0, 0)$, initial point $A(1, 0)$, parameterized counter-clockwise.

Exercise 3.8.4

$\mathbf{v}(x,y) = (x,x)$, C is the unit circle centered at the origin $(0,0)$, initial point $A(1,0)$, parameterized clockwise.

Exercise 3.8.5

$\mathbf{v}(x,y) = (-y,x)$, C is the unit circle centered at the origin $(0,0)$, initial point $A(1,0)$, parameterized counter-clockwise.

Exercise 3.8.6

$\mathbf{v}(x,y) = (-y,x)$, C is the circle of radius $r > 0$ centered at the origin $(0,0)$, initial point $A(1,0)$, parameterized counter-clockwise.

Exercise 3.8.7

$\mathbf{v}(x,y) = (3x,x)$, C is the boundary of the triangle with vertices $P(0,0)$, $Q(10/3,0)$, $R(3,0)$, initial point $P(0,0)$, parameterized counter-clockwise.

Exercise 3.8.8

$\mathbf{v}(x,y) = (3x + y, x + y)$, C is the boundary of the triangle with vertices $P(0,0)$, $Q(10/3,0)$, $R(3,0)$, initial point $P(0,0)$, parameterized counter-clockwise.

Exercise 3.8.9

$\mathbf{v}(x,y) = (12x - 4y, 5x + 7y)$, C is the boundary of the region bounded by the graphs of the functions $y = x^4$ and $y = 16$.

Exercise 3.8.10

$\mathbf{v}(x,y) = (3xy^2, 5x^2y)$, C is the boundary of the triangle with vertices $A(0,0)$, $B(1,0)$, $C(1,1)$, ordered counter-clockwise.

Exercise 3.8.11

$\mathbf{v}(x,y) = (-y,x)$, C is the unit circle centered at the origin $(0,0)$, initial point $A(1,0)$, parameterized clockwise.

Exercise 3.8.12

To generalized Exercise 3.8.5, suppose that $\mathbf{v}(x,y) = (a(x,y), b(x,y))$ satisfies curl $\mathbf{v}(x,y) = 1$. Assuming all integrals are defined and finite, use Green's Theorem to compute the total circulation along ∂R in terms of a quantity associated with the bounded region R.

Exercise 3.8.13

Let $\mathbf{v}(x,y) = \left(\dfrac{-y}{x^2 + y^2}, \dfrac{x}{x^2 + y^2} \right)$ and R be the unit disk centered at $(0,0)$. The circulation of $\mathbf{v}(x,y)$ along ∂R is -2π. But curl $\mathbf{v} = \mathbf{0}$. Does this contradict Green's Theorem? *Hint*: Implicit in the development of Green's Theorem are hypotheses about domains.

Exercise 3.8.14

Let $\mathbf{v}(x,y) = \left(\dfrac{-y}{x^2 + y^2}, \dfrac{x}{x^2 + y^2} \right)$. Compute the circulation along the circle of radius $r > 0$ centered at the origin $(0,0)$.

Exercise 3.8.15

Suppose that $\mathbf{v}(x,y) = (a(x,y), b(x,y)) = \nabla f = (f_x, f_y)$. Show that, for a region R and boundary ∂R, both sides of Green's theorem are zero. Graph and think about streamlines which, in this case, are also called "gradient flows." These are important in physics. For instance, except when there are singularities, air-flow of a weather system follows the gradient of the level curves of air-pressure. The same comment applies to fluid motion, generally.

Exercise 3.8.16

Let $\mathbf{v}(x,y) = \left(\dfrac{-y}{x^2 + y^2}, \dfrac{x}{x^2 + y^2} \right)$ and $f(x,y) = \arctan\left(\dfrac{y}{x} \right)$. Show that

$$\mathbf{v}(x,y) = \nabla f(x,y) = (f_x, f_y) = \left(\frac{\partial f}{\partial x}, \frac{\partial f}{\partial y} \right).$$

For R the unit disk centered at $(0,0)$, compute the circulation along ∂R, counter-clockwise. Does this contradict the previous exercise? Graph and think about streamlines.

3.9   Rotation and circulation density

From the examples, we have accumulating evidence that the algebraic sign of the quantity $b_x - a_y$ correlates with some kind of rotation occurring in streamline trajectories.

However, what is it that rotates? How are to define and compute "rotation" for streamlines?

Let's explore this a bit.

Example

Let $\mathbf{v} = (y, 0)$ be the vector field.

At time $t = 0$, start with reference points $A = (0,1)$ and $B = (0,2)$.

In this case, the formula for the vector field is explicit and linear and so approximation arguments are not needed.

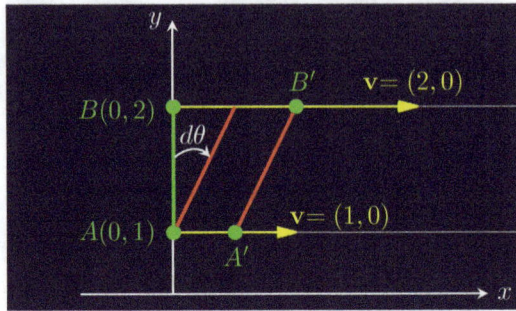

Figure 3.17   Vector field $\mathbf{v} = (y, 0)$. Increment AB flows to increment $A'B'$. Relative change in angle is $d\theta < 0$.

We suppose a small time interval dt. Let's see what happens to $A = (0, 1)$ and $B = (0, 2)$ along their respective integral curves, reaching A' and B', respectively; and then compare AB with $A'B'$.

At $A = (0, 1)$, the velocity vector is $\mathbf{v} = (1, 0)$ and at $B = (0, 2)$, the velocity vector is $\mathbf{v} = (2, 0)$.

So for small time interval dt we get

$$A' = (0, 1) + dt(1, 0) \text{ and } B' = (0, 2) + dt(2, 0).$$

This means that

$$A'B' = (0, 1) + dt(1, 0).$$

But recall that $AB = (0, 1)$. So, for instance, for small $dt > 0$, the corresponding angle differential between AB and $A'B'$ is $d\theta < 0$, clockwise, or negative.

Remember that *circulation density* of $\mathbf{v}$ is $-1 < 0$.

Example

Let $\mathbf{v} = (-y, x)$ be the vector field. At time $t = 0$, start with reference points $A = (1, 1)$ and $B = (2, 2)$. In other words, A and B are along the same radius emanating from the origin $(0, 0)$.

The formula for the vector field is linear and so linear approximation methods are redundant.

We start with the line segment $AB = (1, 1)$ and we suppose a small time interval dt.

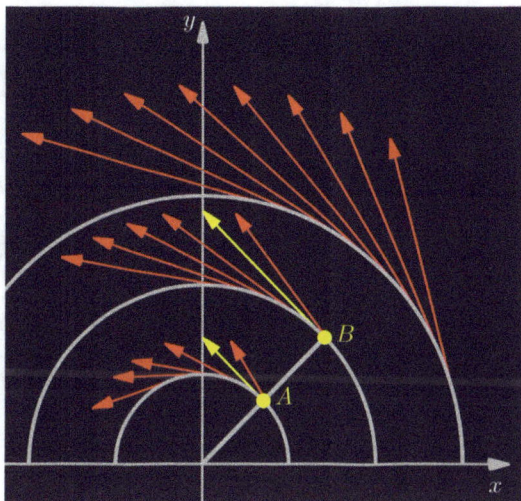

Figure 3.18   Vector field $\mathbf{v} = (-y, x)$. Two initial points are $A = (1, 1)$ and $B = (2, 2)$.

At $A = (1, 1)$, the velocity vector is $\mathbf{v} = (-1, 1)$ and at $B = (2, 2)$, the velocity vector is $\mathbf{v} = (-2, 2)$.

So for small time interval dt we get

$$A' = (1, 1) + dt(-1, 1) \text{ and } B' = (2, 2) + dt(-2, 2).$$

This means that

$$A'B' = (1, 1) + dt(-1, 1).$$

But recall that $AB = (1, 1)$. For small $dt > 0$, the corresponding angle between AB and $A'B'$, is $d\theta$ counterclockwise, or positive.

As is evident from the geometry,

$$\sin(d\theta) = \frac{dt}{1} = dt.$$

For small angle differentials, sine of an angle differential is close to the angle differential (all angles, of course, in radians). Hence, for the small time interval, we get that

$$d\theta \approx \sin(d\theta)dt > 0.$$

We get *circulation density* of $\mathbf{v} = (-y, x)$ is $1 - (-1) = 2 > 0$.

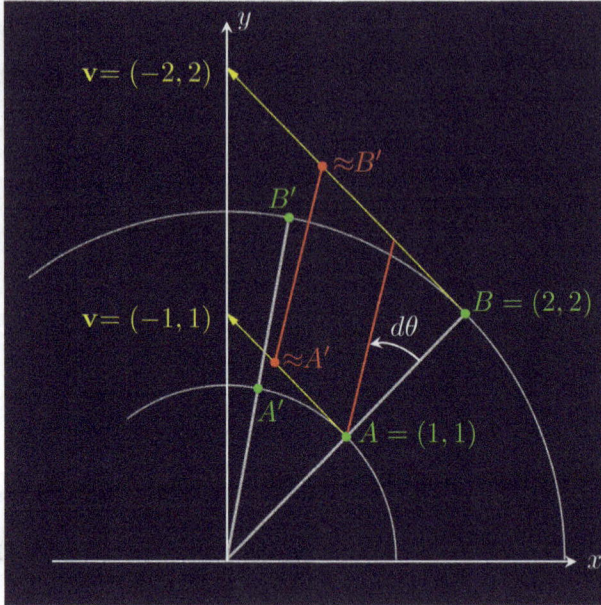

Figure 3.19　Vector field $\mathbf{v} = (-y, x)$. Increment AB flows to increment $A'B'$. Relative change in angle is $d\theta > 0$.

Problem 3.7.

From the geometry $\sin(d\theta) = \dfrac{dt}{1} = dt$.

But for small angle differentials, the sine is close to the angle (all angles in radians). And so, for small time interval, we get that $d\theta \approx dt > 0$.

Evidence continues to accumulate that rotation in some way correlates with circulation density function.

3.10　A general case

Question: In the general case, is there a relationship between $b_x - a_y$ and "rotation" identified by comparing $A'B'$ to AB (where A reaches A' and B reaches B' along their respective streamlines, in a small time interval $dt > 0$)?

More simply expressed, the question can read:

$$[b_x - a_y] \; ? \; [\text{"rotation"} \; AB \; \text{to} \; A'B'].$$

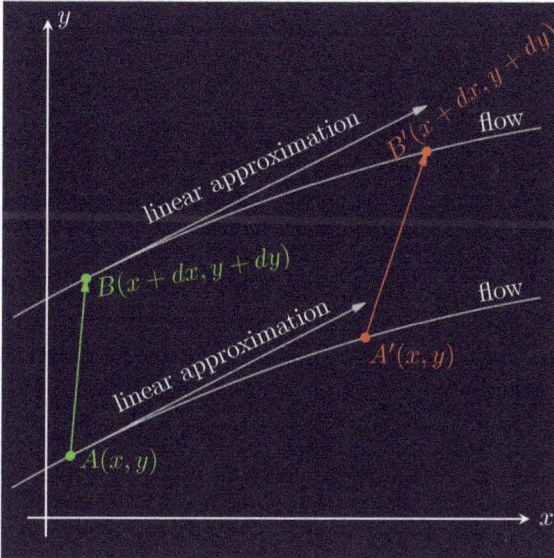

Figure 3.20 A, B, flow to A', B' in small time increment $dt > 0$. In linear approximation, $A'(x,y) \approx A(x,y) + \mathbf{v}(x,y)dt$ and $B'(x + dx, y + dy) \approx B(x + dx, y + dy) + \mathbf{v}(x + dx, y + dy)dt$.

How do we compare two increments? Note that the increments in question are at different locations and their respective coordinates are relative to the coordinate system located at those different locations.

You may recall from Calculus III that comparing increments and tangent vectors obtained at different locations can have its subtleties. The canonical example is the surface of the sphere. The curvature of the sphere needs to be taken into account if one is to suitably "'transport" and compare tangent vectors obtained at different locations.

In our case, all of our computations are on the plane which has zero curvature. This means that we can simply compare coordinates. Implicitly, however, comparing AB and $A'B'$ does require transporting and comparing in a common frame. **With that technicality aside, we can bring precision to our inquiry with linear approximation.**

And so, while several computations follow, they are routine in the sense that we are computing and assembling quantities needed in order to follow through with linear approximation.

The question is about what happens locally, that is, local to a point (x, y). Unlike the examples above, we now suppose that two initial points A and B are relatively close to each other.

And so we write $A = (x, y)$ and $B = (x + dx, y + dy)$, with line segment or "increment" $AB = (dx, dy)$.

For the time being, $AB = (dx, dy)$ is fixed, an initial increment.

In order to compute $A'B'$ (by linear approximation) we need to first find the coordinates of $A'(x, y)$ and $B'(x + dx, y + dy)$.

These are points along respective streamlines of $A(x, y)$ and $B(x + dx, y + dy)$, in small time dt.

Just as in the examples, in order to linearly approximate $A'(x, y)$ and $B'(x + dx, y + dy)$, we need the velocity vectors at both $A(x, y)$ and $B(x + dx, y + dy)$, respectively.

The vector $\mathbf{v}(x, y)$ at $A(x, y)$ is

$$\mathbf{v}(x, y) = \big(a(x, y), b(x, y)\big).$$

The vector $\mathbf{v}(x + dx, y + dy)$ at $B(x + dx, y + dy)$ is

$$\mathbf{v}(x + dx, y + dy) = \big(a(x + dx, y + dy), b(x + dx, y + dy)\big)$$
$$= \big(a(x, y) + a_x dx + a_y dy, b(x, y) + b_x dx + b_y dy\big).$$

To approximate $A'(x, y)$ and $B'(x + dx, y + dy)$ we suppose a small time interval dt. Under the hypotheses of linear approximation

$$A'(x, y) = (x, y) + \big(a(x, y), b(x, y)\big) dt$$

and

$$B'(x, y) = (x + dx, y + dy) + \big(a(x + dx, y + dy), b(x + dx, y + dy)\big) dt$$
$$= (x + dx, y + dy) +$$
$$\big(a(x, y) + a_x dx + a_y dy, b(x, y) + b_x dx + b_y dy\big) dt.$$

Next, we compute and compare the two line segments, AB and $A'B'$.

We have $AB = (dx, dy)$ and from the above computations,

$$A'B' = (dx, dy) + \big(a_x dx + a_y dy, b_x dx + b_y dy\big) dt.$$

Remember that the objective is to compare AB and $A'B'$.

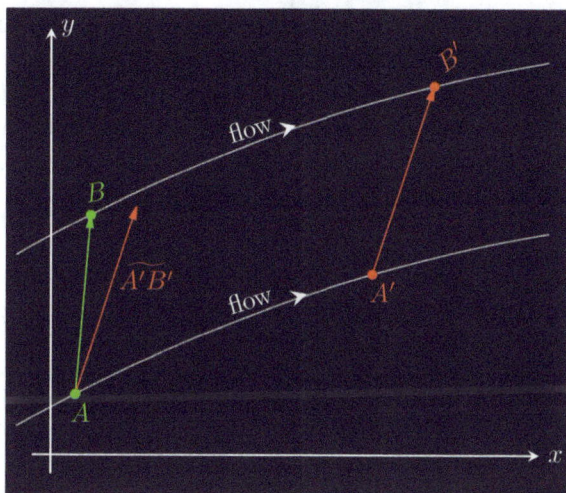

Figure 3.21 $\widetilde{A'B'}$ is congruent and parallel to $A'B'$. Compare AB and $\widetilde{A'B'}$.

The line segment $AB = (dx, dy)$ is a small increment connecting two nearby points A and B. Based on our computations so far, we have that $AB = (dx, dy)$ flows to a new line segment $A'B'$.

It can be helpful to display this in matrix notation:

$$A'B' = \begin{pmatrix} dx \\ dy \end{pmatrix} + \begin{pmatrix} a_x & a_y \\ b_x & b_y \end{pmatrix} \begin{pmatrix} dx \\ dy \end{pmatrix} dt = \left[I + \begin{pmatrix} a_x & a_y \\ b_x & b_y \end{pmatrix} dt \right] \begin{pmatrix} dx \\ dy \end{pmatrix}.$$

Or, with the understanding that $A'B'$ and AB are increments,

$$A'B' = AB + \begin{pmatrix} a_x & a_y \\ b_x & b_y \end{pmatrix} AB dt = \left[I + \begin{pmatrix} a_x & a_y \\ b_x & b_y \end{pmatrix} dt \right] AB$$

where I represents the 2×2 identity matrix.

Alternatively, the *relative strain* is:

$$A'B' - AB = \begin{pmatrix} a_x & a_y \\ b_x & b_y \end{pmatrix} AB dt.$$

The problem is now tractable. The question of relative rotation becomes a precise question of whether or not the linear transformation

$$\mathbf{J v} = \begin{pmatrix} a_x & a_y \\ b_x & b_y \end{pmatrix}$$

has rotational components.

Note that the symbolism '$\mathbf{J v}$' is for the "Jacobian of $\mathbf{v}$", the matrix of first-order derivatives of the vector field $\mathbf{v}(x, y) = \big(a(x, y), b(x, y)\big)$.

We continue to take our lead from classical fluid dynamics

Description of flows reveals two distinct types of motion: "rotational motion," such as when velocities form streamlines that circle a drain; and motion where velocities do not significantly change direction, where motion "is mainly not rotational," such as when water flows along a straight canal.

How might we distinguish these more precisely? We appeal to centuries of successful physics. The Jacobian is, among other things,[31] a linear approximation to the flow velocity field. The question arises, then, as to whether or not we can decompose the Jacobian of the velocity field in a way that corresponds with observable differences in the two types of flow just described.

We will need a result from undergraduate linear algebra, the usefulness of which emerges in the next section.

A matrix $\mathbf{M}$ can be written uniquely as a sum of a *symmetric matrix* $\mathbf{S}$ and an *anti-symmetric* matrix $\mathbf{A}$.

> **Definition 3.3.** A matrix $\mathbf{S}$ is symmetric if and only if $\mathbf{S}^{transpose} = \mathbf{S}$.
> A matrix $\mathbf{A}$ is anti-symmetric if and only if $\mathbf{A}^{transpose} = -\mathbf{A}$.

Note: In this book, all matrices are real. Also, henceforth, the superscript '*transpose*' is abbreviated to '**t**'.

> **Problem 3.8.**
> The decomposition is unique in the sense that if
> $$\mathbf{M} = \mathbf{S} + \mathbf{A}$$
> is a sum of a symmetric and an anti-symmetric matrix, then
> $$\mathbf{S} = \frac{\mathbf{M} + \mathbf{M}^t}{2} \quad \text{and} \quad \mathbf{A} = \frac{\mathbf{M} - \mathbf{M}^t}{2}.$$

Applying this decomposition to the Jacobian, we get
$$\mathbf{Jv} = (\mathbf{Jv}) = (\mathbf{Jv})_S + (\mathbf{Jv})_A = \frac{(\mathbf{Jv}) + (\mathbf{Jv})^t}{2} + \frac{(\mathbf{Jv}) - (\mathbf{Jv})^t}{2},$$
where subscripts S and A have the obvious meanings.

---

[31] It is also the matrix of components of acceleration of the velocity flow. Further aspects of the Jacobian would be studied in a course in fluid dynamics.

Explicitly, this gives

$$\mathbf{Jv} = \begin{pmatrix} a_x & a_y \\ b_x & b_y \end{pmatrix} = \begin{pmatrix} a_x & \frac{a_y + b_x}{2} \\ \frac{a_y + b_x}{2} & b_y \end{pmatrix} + \begin{pmatrix} 0 & -\frac{b_x - a_y}{2} \\ \frac{b_x - a_y}{2} & 0 \end{pmatrix}.$$

Why is this decomposition helpful for our inquiry?

Let's begin with the symmetric part of $\mathbf{Jv}$.

Since

$$(\mathbf{Jv})_S = \frac{(\mathbf{Jv}) + (\mathbf{Jv})^t}{2}$$

is symmetric, there is an orthogonal basis relative to which $(\mathbf{Jv})_S$ can be diagonalized. For the velocity field representing a fluid flow, that basis sometimes is called the basis for the "principle coordinates" of the flow.

Problem 3.9.

In the case of 2×2 matrices, prove the diagonalization result for symmetric matrices. (In higher dimensions, we need the real spectral theorem. That would require more background than is being supposed for this book.)

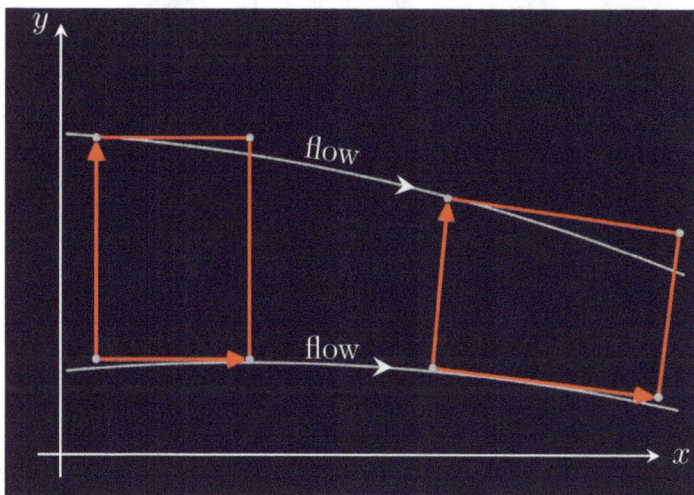

Figure 3.22 Using, principle coordinates of $(\mathbf{Jv})_S$, in linear approximation, rectangles flow to rectangles.

To see why the principle coordinate system is useful, identify a small rectangular area (rectangular relative to the principle coordinate system). If we view the flow of vertices in time relative to these axes (that is, if with a common parameter we view the small rectangle as its vertices flow along their respective streamlines/integral curves) then, while sides may change in length through dilation or contraction relative to the principle axes, for small time increments, the rectangle flows to rectangles.

Dilation and contraction rates are the eigenvalues of $(\mathbf{Jv})_S$. This will become clear in sections to follow.

The anti-symmetric part of $\mathbf{Jv}$ tells us something different.

The anti-symmetric part is

$$(\mathbf{Jv})_A = \frac{(\mathbf{Jv}) - (\mathbf{Jv})^t}{2} = \begin{pmatrix} 0 & -\frac{b_x - a_y}{2} \\ \frac{b_x - a_y}{2} & 0 \end{pmatrix}.$$

Question: Does this type of matrix look familiar?

Example

Suppose that

$$(\mathbf{Jv})_A = \frac{(\mathbf{Jv}) - (\mathbf{Jv})^t}{2} = \begin{pmatrix} 0 & -4 \\ 4 & 0 \end{pmatrix}.$$

Remember that the transformation matrix represents a system of ordinary differential equations for the relative change in increments AB to $A'B'$. To compare, we need a common frame, (u, v) say. Then, by definition, we obtain a system of ordinary differential equations:

$$\begin{cases} \dfrac{du}{dt} = -4v \\[2mm] \dfrac{dv}{dt} = 4u \end{cases}$$

There are methods for solving such systems of equations, studied in undergraduate courses in differential equations. Or, with an ad hoc approach, observe that

$$\mathbf{r}(t) = (x(t), y(t)) = \beta_0(\sin(4t), \cos(4t))$$

solves the initial-value problem $B'(0) = B$, with $\beta_0 = ||B'(0)||$. In other words, the motion is rotation about the origin, in vector coordinates.

Can you now jump to the general case?

If

$$(\mathbf{Jv})_A = \frac{(\mathbf{Jv}) - (\mathbf{Jv})^t}{2} = \begin{pmatrix} 0 & -\frac{b_x - a_y}{2} \\ \frac{b_x - a_y}{2} & 0 \end{pmatrix},$$

then angular velocity of $A'B'$ relative to AB is

$$\frac{b_x - a_y}{2}.$$

This means that we now have a precisely identified geometric meaning for the quantity $b_x - a_y$.

That is,

$$b_x - a_y = 2(\text{angular velocity of } A'B' \text{ relative to } AB).$$

One way to see that is to observe that

$$\mathbf{r}(t) = (x(t), y(t)) = \beta_0 \left(\sin\left(\frac{b_x - a_y}{2}t\right), \cos\left(\frac{b_x - a_y}{2}t\right) \right)$$

is the unique solution to the initial-value problem $B'(0) = B$, with $\beta_0 = \|B'(0)\|$.

For this reason, the anti-symmetric (transformation) matrix

$$(\mathbf{Jv})_A = \frac{(\mathbf{Jv}) - (\mathbf{Jv})^t}{2} = \begin{pmatrix} 0 & -\frac{b_x - a_y}{2} \\ \frac{b_x - a_y}{2} & 0 \end{pmatrix}$$

is also called the *pure rotation* part of the (transformation) matrix

$$\mathbf{Jv} = \begin{pmatrix} a_x & a_y \\ b_x & b_y \end{pmatrix}$$

Summarizing results about $(\mathbf{Jv})_A$ and discovering the name "curl"

Up to first order approximation,

$$\left[A'B' - AB \right]_{\text{due to "pure rotation"}} = \left[\begin{pmatrix} 0 & -\frac{b_x - a_y}{2} \\ \frac{b_x - a_y}{2} & 0 \end{pmatrix} dt \right] AB$$

where

$$(\mathbf{Jv})_A = \frac{(\mathbf{Jv}) - (\mathbf{Jv})^t}{2} = \begin{pmatrix} 0 & -\frac{b_x - a_y}{2} \\ \frac{b_x - a_y}{2} & 0 \end{pmatrix}.$$

Dividing by dt, we obtain

$$\frac{\left[A'B' - AB\right]_{\text{due to "pure rotation"}}}{\text{unit time}} = [\text{rotation matrix}](AB).$$

More formally, let $\mathbf{v}(x, y) = \big(a(x, y), b(x, y)\big)$.

The pure rotation component of the linear approximation of $A'B'$ relative to AB is

$$(\mathbf{Jv})_A = \frac{(\mathbf{Jv}) - (\mathbf{Jv})^t}{2} = \begin{pmatrix} 0 & -\frac{b_x - a_y}{2} \\ \frac{b_x - a_y}{2} & 0 \end{pmatrix}$$

and the term

$$b_x - a_y = 2(\text{angular velocity of } A'B' \text{ relative to } AB).$$

Hence, the term $b_x - a_y$ appropriately is called the (scalar[32]) "***curl***" of the 2-d vector field $\mathbf{v}(x, y) = \big(a(x, y), b(x, y)\big)$. And so we define (scalar)

$$\operatorname{curl} \mathbf{v} := b_x - a_y.$$

Definition 3.4. In fluid dynamics, a 2-d flow is said to be *irrotational* if $\operatorname{curl} \mathbf{v} = b_x - a_y = 0$.

We have solved two problems, but we are getting the same answer in both cases

(1) The quantity $(b_x - b_a)$ is circulation density per unit area.
    (Recall that "circulation" is defined to be a line integral of (components of) mass-flow rates along a curve. In the present context, *circulation density per unit area* is a limit of line integrals, each of which is along a closed curve with (x, y) in the region interior to the curve.)

---

[32]A "vector curl" in 3-d is obtained in Part 2 of this book.

(2) Near the same reference point (x, y) as in 1., the term $(b_x - a_y)$ is twice the angular velocity of $A'B'$ relative to AB. Because of this, it is appropriate to call $(b_x - a_y)$ the (scalar) "curl" of the velocity field $\mathbf{v}(x, y) = \left(a(x, y), b(x, y)\right)$.

Question: Why are the two answers the same? Sorting that out will require digging into subtleties. We will take this up at the end of Part 1.

Let's return to some of our familiar velocity vector fields, to see our results in action.

Example

Let $\mathbf{v} = (-y, x)$. Then $\operatorname{curl} \mathbf{v} = \operatorname{curl}(-y, x) = b_x - a_y = 2$.
Calculating the matrix quantities, we get

$$\mathbf{Jv} = \begin{pmatrix} 0 & -1 \\ 1 & 0 \end{pmatrix}, (\mathbf{Jv})_S = \begin{pmatrix} 0 & 0 \\ 0 & 0 \end{pmatrix}, (\mathbf{Jv})_A = \begin{pmatrix} 0 & 2(-1) \\ 2(1) & 0 \end{pmatrix}.$$

Here,

$$\mathbf{Jv} = \begin{pmatrix} 0 & -1 \\ 1 & 0 \end{pmatrix}$$

is a pure rotation with unit angular velocity, while the angular velocity obtained from the anti-symmetric

$$(\mathbf{Jv})_A = \begin{pmatrix} 0 & 2(-1) \\ 2(1) & 0 \end{pmatrix}$$

is 2.

Example

Let $\mathbf{v} = (x, y)$. This is a vector field where the origin is a source and streamlines are directed along radial arms emanating from the origin. Looking at the graph of the vector field, one might correctly anticipate that relative rotation is minimal or perhaps zero.

Then $\operatorname{curl} \mathbf{v} = \operatorname{curl}(x, y) = b_x - a_y = 1 - 1 = 0$. Calculating the matrix quantities, we get

$$\mathbf{Jv} = \begin{pmatrix} 1 & 0 \\ 0 & 1 \end{pmatrix}, (\mathbf{Jv})_S = \begin{pmatrix} 1 & 0 \\ 0 & 1 \end{pmatrix}, (\mathbf{Jv})_A = \begin{pmatrix} 0 & 0 \\ 0 & 0 \end{pmatrix}.$$

For this vector field, up to first order approximation, relatively close streamlines are parallel and in sync.

Exercises

Exercise 3.10.1
Let $\mathbf{v}(x, y) = (1, 1)$.

(a) Graph the velocity field and its streamlines.

(b) Find explicit formulas for the streamlines, in the form $\mathbf{r}(u) = (x(u), y(u))$, where u is a real variable.

(c) Compute curl $\mathbf{v}(x, y)$. Explain.

Exercise 3.10.2
Let $\mathbf{v}(x, y) = (x, x)$.

(a) Graph the velocity field and its streamlines.

(b) Set up appropriate differential equations and find explicit formulas for the streamlines, in the form $\mathbf{r}(u) = (x(u), y(u))$, where u is a real variable.

(c) Compute curl $\mathbf{v}(x, y)$.

(d) The streamlines are parallel but there is a non-zero rotational component. Explain. *Hint*: You might start by comparing $\mathbf{v}(1, 0)$ and $\mathbf{v}(0, 1)$. Remember that the curl $\mathbf{v}(x, y)$ is a first-order approximation.

(e) Compute $(\mathbf{Jv})_S$ and $(\mathbf{Jv})_A$.

Exercise 3.10.3
Let $\mathbf{v}(x, y) = (x, y)$.

(a) Graph the velocity field and its streamlines.

(b) Set up appropriate differential equations and find explicit formulas for the streamlines, in the form $\mathbf{r}(u) = (x(u), y(u))$, where u is a real variable.

(c) Compute curl $\mathbf{v}(x, y)$.

(d) The streamlines are *not* parallel but the rotational component is zero. Explain. *Hint*: You might start by comparing $\mathbf{v}(1, 0)$ and $\mathbf{v}(0, 1)$. Remember that the curl $\mathbf{v}(x, y)$ is a first-order approximation.

(e) Compute $(\mathbf{Jv})_S$ and $(\mathbf{Jv})_A$.

Chapter 4

2-D FLUX AND DIVERGENCE

4.1 2-dimensional flux

As in previous sections, suppose $\mathbf{v}(x,y) = \big(a(x,y), b(x,y)\big)$ is a 2-component vector field defined in the plane. Green's theorem emerges from inquiring about mass-flow rate along a curve which, for consistency, is required to be positively oriented. (If we look to a small rectangle $dx \times dy$, linear approximation gives that the mass-flow rate along the positively oriented boundary of the rectangle is $(b_x - a_y) \times dxdy$.)

We now look to a complementary problem that also is inspired by physics.

Question: What is the *net* mass-flow rate *across* a small rectangular region, in other words, what is the "mass-flow rate out - mass-flow rate in"?

As in previous sections, the small rectangle $dx \times dy$ has four sides, **I, II, III** and **IV**.

We estimate "mass-flow rate out minus mass-flow rate in."

To do that, we use the components of $\mathbf{v} = (a,b)$ that are orthogonal to each side, and subtract appropriately.

That is, for the small rectangle, the *net* mass-flow rate is

[Net rate in the y-direction] + [net rate in the x-direction]

$= $ [rate across **III** – rate across **I**] + [rate across **II** – rate across **IV**]

$= [b(x + dx, y + dy)dx - b(x,y)dx] + [a(x + dx, y)dy - a(x, y + dy)dy].$

This is similar to the computation for mass-flow rate along a curve which, as you may recall, was a line integral.

Question: But is this last equation a line integral?

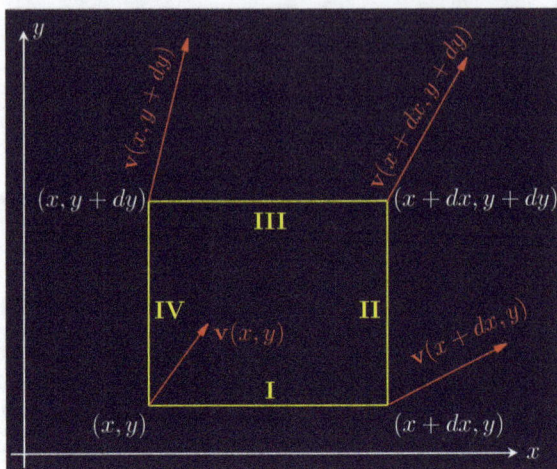

Figure 4.1 Net mass-flow rate through region $dx \times dy$ is net mass-flow rate in y-direction minus net mass-flow rate in x-direction. Use normal components to sides **I**, **II**, **III** and **IV**, respectively. That is, mass-flow rate is $[b(x + dx, y + dy)dx - b(x, y)dx] + [a(x + dx, y)dx - a(x, y + dy)dx]$.

Clue: Factors can be obtained by identifying components of the vector field that are orthogonal to sides **I**, **II**, **III** and **IV**, as the case may be.

Problem to solve: The computation so far is ad hoc, obtained by appealing to a diagram for a region rather than by appealing to a defined line integral.

To see how we might rewrite this (physically significant quantity) as a line integral, let's begin by organizing terms so that, as hinted at in the clue above, sides are in the right order to potentially give a positively oriented boundary curve for the rectangle $dx \times dy$.

That is, we order the sides as follows:

I followed by **II**, **II** followed by **III** and then **III** followed by **IV**, and so **IV** back to the initial point.

By looking at the figure, the net mass-flow rate across the small region is, then,

$$-b(x, y)dx + a(x + dx, y)dy + b(x + dx, y + dy)dx - a(x, y + dy)dy.$$

Think about geometry and vectors. In particular, remember the clue given above, regarding orthogonality.

The first term $-b(x, y)$ is the dot product of $\mathbf{v}(x, y) = \Big(a(x, y), b(x, y)\Big)$ with the vector $\mathbf{N} = (0, -1)$ that is normal to side **I** and pointing downward.

The second term $a(x + dx, y)$ is the dot product of $\mathbf{v}(x, y)$ with the vector $\mathbf{N} = (1, 0)$ that is normal to side **II** and pointing to the right in the positive x-direction. And so on.

Problem 4.1.
Provide a similar identification for the other two terms.

Observation: In each case we get the right terms, so long as we compute the projection (or dot product) onto what, relative to the positively oriented boundary is the "outward directed unit normal vector **N**."

Therefore, the net mass-flow rate across the region $dx \times dy$ can indeed be written as a line integral in the following way:

The rectangle has sides **I**, **II**, **III** and **IV**. Link them up to obtain a positively oriented boundary curve for the region

$$C = \mathbf{I} + \mathbf{II} + \mathbf{III} + \mathbf{IV},$$

where addition of curves is defined in Section 3.6.

Let $\mathbf{N}$ be the outward directed unit normal vector along each side. Then an approximation to the net mass-flow rate through the small rectangular region is

$$\oint_C \mathbf{v} \cdot \mathbf{N} \, ds.$$

A technical issue regarding the integral for net mass-flow rate across a small rectangular region

Can the integral that we just defined be expressed as a line integral of the form

$$I = \int_C f \, dx + g \, dy,$$

where f and g are functions defined on the plane?

To compute

$$\oint_C \mathbf{v} \cdot \mathbf{N} \, ds,$$

we need terms for the integrand.

These are

$$\mathbf{v} = (a, b), \quad \mathbf{N} = \frac{(y', -x')}{\| (y', -x') \|}, \quad ds = \| (x', y') \| \, du, \quad dx = x' du, \quad dy = y' du.$$

Observe that $\| (y', -x') \| = \| (x', y') \|$.

Substituting, we get

$$\oint_C \mathbf{v} \cdot \mathbf{N} \, ds = \int_C (a, b) \cdot (dy, -dx) = \int_C -b \, dx + a \, dy.$$

In other words, the net mass-flow rate is a line integral of the form

$$I = \int_C f \, dx + g \, dy.$$

That is, by setting $f = -b$ and $g = a$, we get

$$\oint_C \mathbf{v} \cdot \mathbf{N} ds = \int_C f \, dx + g \, dy.$$

Problem 4.2.
The 2-d flux integral is invariant under change of parameterization. Provide details.

Continuing with the main discussion

For the small region, the net mass-flow rate is

[rate across **III** − rate across **I**] + [rate across **II** − rate across **IV**]

$$= [b(x + dx, y + dy)dx - b(x, y)dx] + [a(x + dx, y)dy - a(x, y + dy)dy].$$

Problem 4.3.
Just as we did in our inquiry that led to Green's theorem, by linear approximation (ignoring high order difference terms $dx dx$, $dy dy$, $dx dx dy$, $dy dy dx$, $\cdots$ which in the limit do not contribute to integrals), this combination of mass-flow rates terms can be written as

[rate across **III** − rate across **I**]+[rate across **II** − rate across **IV**]

$$= [a_x + b_y]dx dy.$$

Consequently, near a point (x, y), the net mass-flow rate through the small rectangular region $dx \times dy$ is $(a_x + b_y)(dx dy)$.

Dividing by the area term, we get that the net mass-flow rate per unit area is $a_x + b_y$.

Just as in Green's theorem, we obtain a density function which in this case is:

$$[\text{flux density (per unit area) of } \mathbf{v} = (a, b)] = [a_x + b_y].$$

The quantity $a_x + b_y$ is also called the *divergence* of the vector field and is often written as div $\mathbf{v}$.

Just as in the development of Green's theorem, by linear approximation we get two ways of approximating the same quantity. And the resulting theorem can be left as an exercise.

> **Problem 4.4. (2-d divergence theorem)**
> Provide the main details and, with appropriate hypotheses in place (e.g., that the boundary curve ∂R of the region R be positively oriented), show that having the two ways of representing the same quantity leads to the following integral formula:
>
> $$\oint_{\partial R} -b\,dx + a\,dy = \int_R (a_x + b_y)\,dx\,dy.$$

If we include other standard ways of writing these integrals, we get the following "line up":

$$\oint_C \mathbf{v} \cdot \mathbf{N}\,ds = \int_C (a, b) \cdot (dy, -dx)$$

$$= \int_C -b\,dx + a\,dy$$

$$= \oint_{\partial R} -b\,dx + a\,dy$$

$$= \int_R (a_x + b_y)\,dx\,dy$$

In other words, under suitable hypotheses, for a vector field $\mathbf{v}$, the flux across the boundary of the region is equal to the integral of the divergence over the interior of the region.

Note: This theorem is a 2-d analogue for the classical divergence theorem in 3-d that 'in physics' also is called Gauss' theorem.

Exercises

In Exercises 4.1.1–4.1.7, for each vector field $\mathbf{v}(x, y) = (a(x, y), b(x, y))$, where possible, compute the flux across the boundary ∂R of the region R, oriented counter-clockwise: (i) directly by integrating along ∂R; and (ii) by using the 2-d divergence theorem (Problem 4.4). Graph the vector field and its streamlines and, if possible, find explicit formulas for the streamlines by solving an appropriate system of differential equations. Graph and explain your results.

Exercise 4.1.1

$\mathbf{v}(x,y) = (1,0)$, R is the square with vertices $A(0,0)$, $B(1,0)$, $C(1,1)$, $D(0,1)$.

Exercise 4.1.2

$\mathbf{v}(x,y) = (1,0)$, R is the triangle with vertices $A(0,0)$, $B(1,1)$, $C(0,1)$.

Exercise 4.1.3

$\mathbf{v}(x,y) = (x,y)$, R is the unit circle centered at the origin.

Exercise 4.1.4

$\mathbf{v}(x,y) = (1,1)$, R is the unit circle centered at the origin.

Exercise 4.1.5

$\mathbf{v}(x,y) = (x,x)$, R is the square with vertices $A(0,0)$, $B(1,0)$, $C(1,1)$, $D(0,1)$.

Exercise 4.1.6

$\mathbf{v}(x,y) = (x,x)$, R is the square with vertices $A(1,0)$, $B(2,1)$, $C(1,2)$, $D(0,1)$.

Exercise 4.1.7

$\mathbf{v}(x,y) = (-y,x)$, R is the unit circle centered at the origin.

Exercise 4.1.8

Let $\mathbf{v}(x,y) = \left(\dfrac{-y}{x^2 + y^2}, \dfrac{x}{x^2 + y^2} \right)$ and R be the unit disk centered at $(0,0)$. Show that both sides of the divergence theorem are zero. But why does this not provide an illustration of the divergence theorem? *Hint*: Consider the domains.

Exercise 4.1.9

This exercise relates to Exercises 4.1.3–4.1.4:

$$\mathbf{v}(x,y) = \nabla f(x,y) = (f_x, f_y) = \left(\frac{\partial f}{\partial x}, \frac{\partial f}{\partial y} \right),$$

R a region whose boundary ∂R is given by a simple closed curve $\mathbf{r}(u) = (x(u), y(u))$, parameterized to track counter-clockwise. Show that

$$\iint_R \operatorname{div} \mathbf{v} \, dxdy = \iint_R (a_x + b_y) \, dxdy = \int_{\partial R} -b \, dx + a \, dy = 0.$$

4.2 Complementarity of 2-d flux and 2-d circulation

If you are getting comfortable handling line integrals, perhaps you have spotted that we can get the 2-d divergence theorem from Green's theorem, without having to argue from first principles. Although, looking for that alternate approach usually presupposes some prior grasp of the divergence theorem. For then one is on the lookout for a derivation of what one already knows within the axiomatic system.

Problem 4.5. (The 2-d divergence theorem from Green's theorem; and vice versa)
We have two integral formulas:

Green's theorem: $\displaystyle\oint_{\partial R} a \, dx + b \, dy = \int_R (b_x - a_y) \, dxdy$

and the

2-d divergence theorem: $\displaystyle\oint_{\partial R} -b \, dx + a \, dy = \int_R (a_x + b_y) \, dxdy.$

The formulas are obtained by answering different questions about a vector field $\mathbf{v} = (a, b)$. The questions and answers are different. At the same time, there are evident similarities. Are the two formulas related? Or, in symbols,

$$\left[\oint_{\partial R} a \, dx + b \, dy = \int_R (b_x - a_y) \, dxdy\right]$$

$$?$$

$$\left[\oint_{\partial R} -b \, dx + a \, dy = \int_R (a_x + b_y) \, dxdy\right]$$

Clue: Under our hypotheses, the vector field $\mathbf{v} = (a, b)$ was arbitrary.

To follow up on that clue, let's use different symbols for each formula.

Rewriting, we get

Green's theorem:

$$\oint_{\partial R} a \, dx + b \, dy = \int_R (b_x - a_y) \, dxdy$$

and

2-d divergence theorem:

$$\oint_{\partial R} -f\,dx + e\,dy = \int_R (e_x + f_y)\,dxdy.$$

Evidently, a linkage between the formulas is established by setting $-f = a$, $e = b$.

That is, given the ordered pair (a, b), construct the new ordered pair $(b, -a)$.

> **Problem 4.6.** Show that this substitution allows us to derive Green's Theorem from the 2-d divergence theorem; and vice versa.

Question: What is the geometry of this linkage?

The vector field $\mathbf{w} = (b, -a)$ is everywhere orthogonal to $\mathbf{v} = (a, b)$.

Hence, a circulation integral of $\mathbf{v} = (a, b)$ along a curve is a flux integral of $\mathbf{w} = \mathbf{v}_\perp = (b, -a)$ along the same curve; and vice versa.

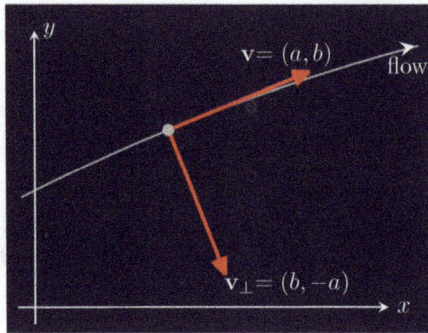

Figure 4.2   For $\mathbf{v} = (a, b)$, $\mathbf{v}_\perp = (b, -a)$.

Example

For $\mathbf{v} = (2, 0)$, $\operatorname{div}\mathbf{v} = \operatorname{div}(2, 0) = 0 + 0 = 0$.

Example

For $\mathbf{v} = (y, 0)$, there is a source term in the first component that increases with y.

The divergence is $\operatorname{div}(y, 0) = 1 + 0 = 1$. In other words, as y increases, so does the divergence.

Example

Let $\mathbf{v} = (-y, x)$ be the familiar vector field whose streamlines orbit the origin $(0,0)$. As you might expect, the divergence is zero. Explicitly, div $(-y, x) = 0 + 0$.

Example

Imagine a round fountain with a conical surface, and a steady source of water that flows from its top. Water flows down the conical surface in all directions. It has (approximately) constant acceleration due to gravity until at the bottom of the cone's edge, water falls into a surrounding channel that collects water for recirculation through the fountain.

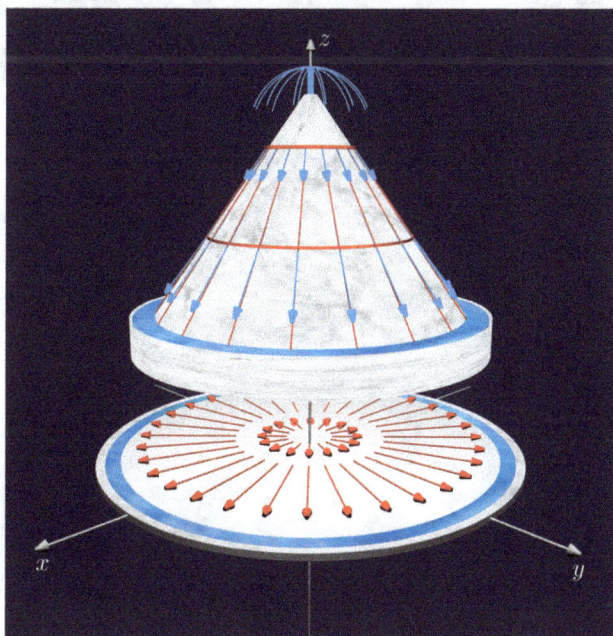

Figure 4.3   Water fountain and aerial view of water-flow velocity field.

Problem 4.7. Explain why the depth of water decreases as it spreads out along the surface of the cone. *Hints*: Polar coordinates, increments in radius, and water volume. From physics, assume that water is not destroyed.

With an aerial view (that is, focusing on x and y coordinates and ignoring height of cone and total depth of water at locations along the surface of the cone), streamlines flow radially, away from the origin; and the magnitude of the vectors (speed) increase with distance from the origin.

A simplistic mathematical model that has similar features is $\mathbf{v}(x, y) = (x, y)$.

For this vector field, divergence is $\operatorname{div}(x, y) = 1 + 1 = 2$.

The flux density per unit area is 2.

Let's review what this means in this case.

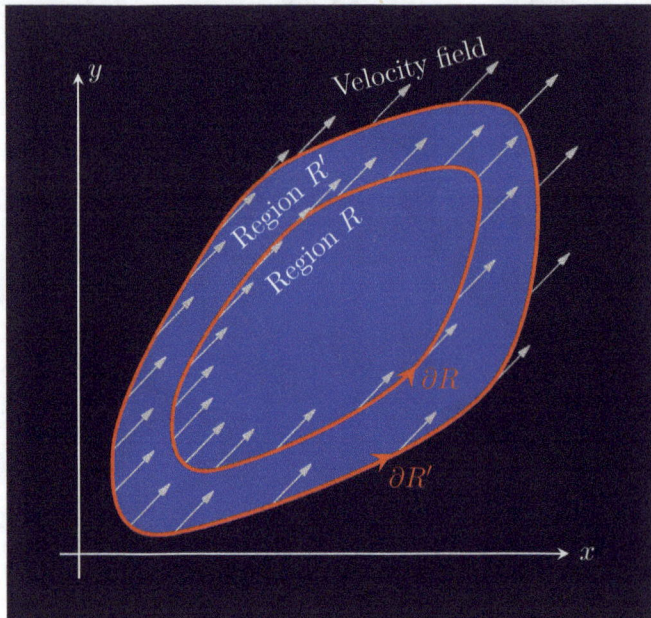

Figure 4.4   Approximating difference in flux using difference area.

Start with a reference point (x, y) and suppose that $C = \partial R$ is a positively oriented closed curve that is boundary to a small region R containing the point (x, y). Replace R by a slightly (linear approximation) larger area R', one that includes R (see Figure 4.4).

There is a difference in area dA.

What is $d(flux)$, the difference in the flux integral for $C' = \partial R'$?

Since $\operatorname{div}(x, y) = 1 + 1 = 2$ is flux density per unit area, $d(flux)$ approximately $2dA$.

Discovering that the quantity div v $= a_x + b_y$ is also a "rate of relative change in area"[33]

The 2-d divergence theorem relates flux along a closed curve (whose interior contains a fixed reference point) to a density function per unit area. But as can be observed in fluid motion, there is also motion of "elements of area."

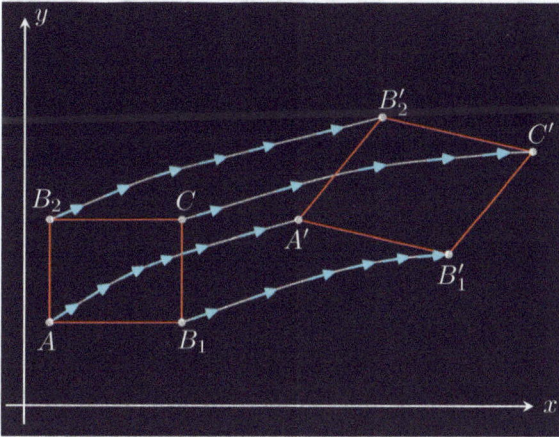

Figure 4.5   Vertices AB_1CB_2 flow to $A'B_1'C'B_2'$.

We can, for instance, imagine test particles at vertices of a rectangle. As some examples already reveal, in some cases "surface areas of rectangles expand" while, in other cases, surface areas decrease (for instance, when a flow is forced into a narrow channel).

This leads to the following question:

Question: For a given velocity vector field, how do areas change as vertices move along their respective streamlines?

Remark: Are you noticing a similarity here with the line of questioning that led to a second derivation of the circulation density function, $b_x - a_x$? In that case, a question came up about the *relative change of increments*, $A'B'$ and AB, in time. The present question is similar. Now, however, we ask about *relative change in areas*, in time.

[33]In fluid dynamics, this is called "strain" in 2-d motion.

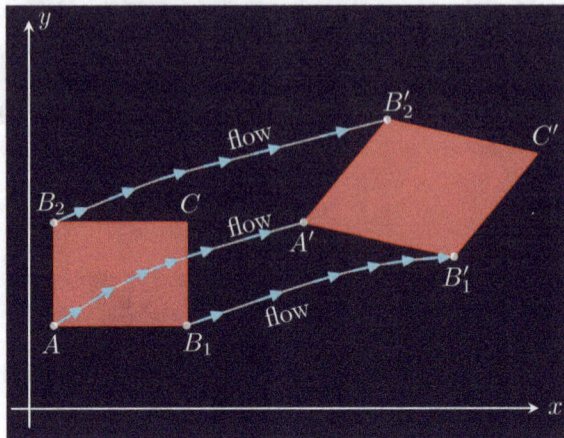

Figure 4.6   Rectangle determined by Vertices A, B_1, B_2 yields rectangle determined by A', B_1', B_2'.

Preliminaries: Three reference points A, B_1 and B_2 determine a parallelogram with sides AB_1 and AB_2 and area $[Area]$. (They might be collinear but draw the picture as though they are not.) If we track the motions of these points along their respective streamlines to A', B_1' and B_2' , we get a new parallelogram, one with sides $A'B_1'$ and $A'B_2'$ and area $[Area]'$.

The question now is to compare the two areas,

$$[Area]' - [Area],$$

as a function of time. This means that we need to use a common integral curve parameter for all three streamlines.

Warmup problems

Warmup 1:
Let $\mathbf{v}(x, y) = (x^2, 0)$. In this example, there is acceleration in the first component of the velocity field, relative to change in the first coordinate. Let $A = (1, 0)$, $B_1(2, 0)$ and $B_2(1, 1)$. In this case, initial $[Area] = 1$.

Motion is in the direction of the x-axis. Suppose that the initial value of the parameter for the integral curves is $t = 0$ and that time increases.

At $t = 0$, $\mathbf{v}(1, 0) = (1, 0)$, $\mathbf{v}(2, 0) = (4, 0)$, and $\mathbf{v}(1, 1) = (1, 0)$.

For a small time increment $dt > 0$, $A = (1, 0)$ and $B_2 = (1, 1)$ both advance in the x direction by 1.

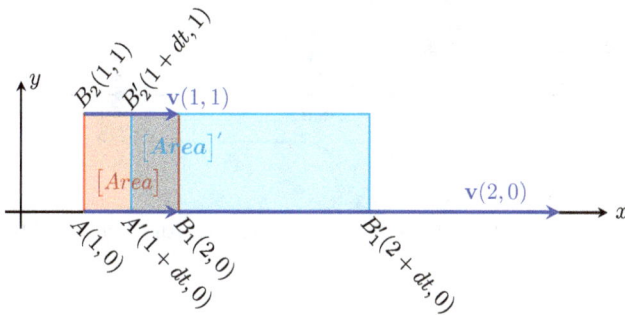

Figure 4.7 $[Area]' - [Area] = (1 + 3dt) - 1 = 3dt.$

And so $A' = (1 + dt, 0)$ and $B_2' = (1 + dt, 1)$. But at $\mathbf{v}(2, 0) = (4, 0)$. This means that B_1 advances to $B_1' = (2 + 4dt, 0)$ (in linear approximation). The original square has been both translated and dilated in the x direction. Explicitly,

$$[Area]' = [(2 + 4dt) - (1 + dt)][1 - 0]$$
$$= 1 + 3dt.$$

Therefore,

$$[Area]' - [Area] = (1 + 3dt) - 1 = 3dt.$$

In other words, the area is an increasing function.

Warmup 2:
In the previous example, points were separated by integer lengths. And, the change in areas is completely determined by the first coordinate function of the vector field, namely, $a(x, y) = x^2$.

Before tackling the general problem, let's try a generic one-dimensional problem, $v = v(x)$.

In this case, the question is about how interval lengths L change in time along the x-axis.

We suppose a small time increment $dt > 0$.

Initial points are $A = x$ and $B = x + dx$, with $L = dx$. The increment $dx > 0$ is fixed.

The velocity (vectors) at these points are $v(x)$ and

$$v(x + dx) \approx v(x) + v_x dx.$$

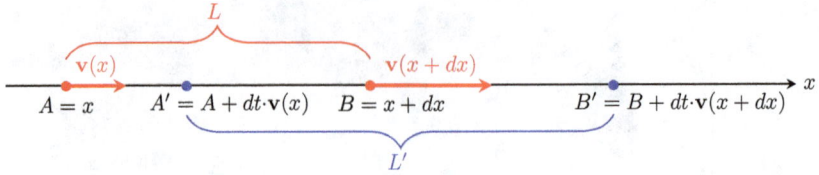

Figure 4.8　A flows to A', B flows to B'. Compare lengths of $L = ||AB||$ and $L' = ||A'B'||$.

By linear approximation, in a small time increment $dt > 0$, these points flow to

$$A' = x + v(x)dt$$

and

$$B' = x + dx + v(x + dx)dt.$$

Because

$$v(x + dx) \approx v(x) + v_x dx,$$

these are

$$A' = x + v(x)dt$$

and

$$B' = x + dx + [v(x) + v_x dx]dt.$$

This gives

$$L' = dx + [v_x dx]dt.$$

But $L = dx$, which means that

$$L' - L = [v_x dx]dt. \tag{4.1}$$

Dividing by dt we get

$$\frac{L' - L}{dt} = [v_x L].$$

What does this mean?

For the argument, the increment $L = dx > 0$ is fixed.

It tells us that as we follow a path of intervals whose end points are determined by integration of the one-dimensional vector field, the lengths of the intervals change in time at a rate of $v_x dx$, namely, the one-dimensional divergence of the velocity field $v = v(x)$ times the length of the initial interval $L = dx > 0$.

What happens if $dx = 0$? The right hand side of equation 4.1 is zero because then $[v_x dx] = [v_x \cdot 0] = 0$.

What about the left side of the equation? What is happening in the geometry?

When $dx = 0$, the initial interval is of length zero. Then $A = B$ and $A' = B'$, which means that all lengths $L = 0 = L'$.

4.3   Divergence in 2-d and rate of relative change in areas

Let's now work out the general case in 2 dimensions.

Suppose that $\mathbf{v}(x, y) = \Big(a(x, y), b(x, y)\Big)$ is a velocity vector field in the plane. As above, we start with three initial points A, B_1 and B_2, where $A = (x, y)$, $B_1 = (x + dx, y)$ and $B_2 = (x, y + dy)$.

These three points determine a parallelogram (in fact, a rectangle lined up with the axes) with sides AB_1 and AB_2 with area $[Area] = dx dy$.

Suppose a common small time increment $dt > 0$.

Using that common parameter, we approximate new locations along streamlines, A', B_1' and B_2'.

These three points also determine a parallelogram, with area $[Area]'$. See Figure 4.6. The problem is to compute

$$\frac{[Area]' - [Area]}{dt}.$$

In order to approximate A', B_1' and B_2', we first need to know the velocity vectors at each of $A = (x, y)$, $B_1 = (x + dx, y)$ and $B_2 = (x, y + dy)$.
At $A = (x, y)$,

$$\mathbf{v}(x, y) = \Big(a(x, y), b(x, y)\Big).$$

At $B_1 = (x + dx, y)$,

$$\mathbf{v}(x + dx, y) = \Big(a(x + dx, y), b(x + dx, y)\Big)$$
$$= \Big(a(x, y) + a_x dx, b(x, y) + b_x dx\Big).$$

At $B_2 = (x, y + dy)$,

$$\mathbf{v}(x, y + dy) = \Big(a(x, y + dy), b(x, y + dy)\Big)$$
$$= \Big(a(x, y) + a_y dy, b(x, y) + b_y dy\Big).$$

Therefore,

$$A' = (x, y) + (a, b)dt$$

and

$$B_1' = (x + dx, y) + \mathbf{v}(x + dx, y)dt$$
$$= (x + dx, y) + \Big(a(x, y) + a_x dx, b(x, y) + b_x dx\Big)dt.$$

Also,

$$B_2' = (x, y + dy) + \mathbf{v}(x, y + dy)dt$$
$$= (x, y + dy) + \Big(a(x, y) + a_y dy, b(x, y) + b_y dy\Big)dt.$$

We can now compute the sides of the new parallelogram:

$$A'B_1' = (dx, 0) + (a_x dx, b_x dx)dt,$$

and

$$A'B_2' = (0, dy) + (a_y dy, b_y dy)dt.$$

Writing the pair of vectors in matrix form,

$$\begin{pmatrix} A'B_1' \\ A'B_2' \end{pmatrix} = \begin{pmatrix} (1 + a_x dt)dx & b_x dtdx \\ a_y dtdy & (1 + b_y dt)dy \end{pmatrix}$$

Problem 4.8.

Using the absolute value of the determinant, compute the area determined by $A'B_1'$ and $A'B_2'$. Show that

$$[Area]' = |(1 + a_x dt)dx(1 + b_y dt)dy - (a_y dtdy)(b_x dtdx)|.$$

This reduces to

$$|(1 + a_x dt + b_y dt)dxdy + (\text{higher order error terms})dxdy|.$$

For a discussion about why the determinant function provides *area*, see Section 1.7.

Recall that $[Area] = dxdy$. And so, up to linear approximation,

$$\frac{[Area]' - [Area]}{dt} = (a_x + b_y)dxdy.$$

That is,

$$\frac{[Area]' - [Area]}{dt} = (a_x + b_y).$$

Note: This is not a rate of change of area. The solution is a rate of change of *relative change in area*, determined by the flow of vertices along streamlines (integral curves) that are simultaneously parameterized.[34]

Definition 4.1. In fluid dynamics, if the rate of relative change of area is zero then mass is neither destroyed nor created. For that reason, a flow for which

$$\nabla \cdot \mathbf{v} = (a_x + b_y) = 0$$

is called *incompressible*.

Concluding remarks regarding relative change in area

Start with a small element of area and examine its progression for small time, relative to the original small element of area. Consider the relative change of area, with respect to time. In the limit, we obtain $a_x + b_y$.

Recall that $a_x + b_y$ is called the divergence of the vector field, where that name originates from estimating net mass-flow rates (flux) through a small region whose boundary is a positively oriented curve.

Just as in our inquiry about circulation, two different questions have the same answer:

At a point (x, y):

(1) The flux density per unit area is
$$\text{div } \mathbf{v} = \text{div }(a, b) = a_x + b_y.$$

(2) The time derivative of relative change in area is
$$\text{div } \mathbf{v} = \text{div }(a, b) = a_x + b_y.$$

---

[34]See note 8.

Question: Why are these answers the same? In classical fluid dynamics, this is equivalent to a conservation law for a type of fluid. In Chapter 5, we go into some of the mathematical details.

Summary of results about "curl" and "divergence" in 2-d

Suppose a velocity vector field $\mathbf{v}(x,y) = \Big(a(x,y), b(x,y)\Big)$.

The Jacobian of the vector field is

$$\mathbf{Jv} = \begin{pmatrix} a_x & a_y \\ b_x & b_y \end{pmatrix}.$$

The Jacobian can be written uniquely as a sum of symmetric and anti-symmetric parts. That is,

$$\mathbf{Jv} = \big(\mathbf{Jv}\big)_S + \big(\mathbf{Jv}\big)_A$$

where the symmetric part is

$$\big(\mathbf{Jv}\big)_S = \begin{pmatrix} a_x & \frac{b_x + a_y}{2} \\ \frac{b_x + a_y}{2} & b_y \end{pmatrix}$$

and the ant-symmetric part is

$$\big(\mathbf{Jv}\big)_A = \begin{pmatrix} 0 & -\frac{b_x - a_y}{2} \\ \frac{b_x - a_y}{2} & 0 \end{pmatrix}.$$

At a point (x,y):

Circulation:

(1) At a fixed time t, using a rectangle $dx \times dy$, the *circulation density function per unit area* is $b_x - a_a$. (By linear approximation)

(2) By summing both sides of (1), and taking limits as partitions are refined, we get **Green's Theorem** in the plane, that is:

$$\int_{\partial R} a\,dx + b\,dy = \int_R (b_x - a_y)\,dx\,dy.$$

Or, in words, with modest hypotheses, the circulation of $\mathbf{v} = (a, b)$ along the closed boundary of R is equal to the area integral of curl $\mathbf{v}$ over the region R.

(3) Following increments along streamlines, the quantity $(b_x - a_y)$ is twice the angular velocity of $A'B'$ relative to AB (using a common integral curve parameter). For that reason, $(b_x - a_y)$ is called the "**curl**" of the vector field.

(4) In terms of the Jacobian $\mathbf{Jv}$, curl is a difference of off-diagonal elements. Alternatively, it is the off-diagonal element of the pure-rotational (anti-symmetric) component $(\mathbf{Jv})_A$ of the Jacobian $\mathbf{Jv}$.

Flux:

(1) At a fixed time t, using a rectangle $dx \times dy$, the *flux density function per unit area* is div $\mathbf{v}$ = div $(a, b) = a_x + b_y$. (Linear approximation)

(2) By summing both sides of (1), and taking limits as partitions are refined, we get the **2-d divergence theorem**, that is:

$$\int_{\partial R} -b\,dx + a\,dy = \int_R (a_x + b_y)\,dx\,dy.$$

Or, in words, with modest hypotheses, the flux of $\mathbf{v} = (a, b)$ along the closed boundary of R is equal to the area integral of div $\mathbf{v}$, over the region R.

(3) Following areas as vertices flow along streamlines, the rate of relative change in area is div $\mathbf{v}$ = div $(a, b) = a_x + b_y$ (using common integral curve parameter).

(4) In terms of the Jacobian $\mathbf{Jv}$, the divergence

$$\text{div}\,(a, b) = a_x + b_y = \text{trace}(\mathbf{Jv}) = \text{trace}(\mathbf{Jv})_S.$$

Chapter 5

UNIFYING 2-D RESULTS

5.1 Unifying results

The following discussion is for readers with some background in undergraduate linear algebra.

In the summary given in Chapter 4, similarities and linkages invite further inquiry.

(1) Green's theorem and the 2-d divergence theorem tell us different things about a vector field. Mathematically, however, the two theorems are equivalent.

(2) In Green's theorem, circulation density $(b_x - a_y)$ is also equal to

$$2[\text{angular velocity of increments } A'B' \text{ relative to } AB]$$

(as defined above).

(3) In the 2-d divergence theorem, flux density $a_x + b_y$ is also equal to the rate of relative change in areas (as defined above).

(4) The flux density is trace $(\mathbf{Jv})_S$; and circulation density is "signed off-diagonal trace" of $(\mathbf{Jv})_A$.

What would be nice is a unified account of these similarities and linkages.

Writing both flux density and circulation density in terms of Jacobians

Flux density is trace $(\mathbf{Jv})_S$ = trace $(\mathbf{Jv})$. Might *circulation density* $b_x - a_y$ be written in a similar way?

Circulation density is an "off-diagonal trace" of $(\mathbf{Jv})_A(x, y)$. Like flux density, is circulation density also the trace of some matrix?

Certainly, it is, if we "manually" construct the new matrix

$$\begin{pmatrix} b_x & b_y \\ -a_x & -a_y \end{pmatrix}.$$

That is,

$$b_x - a_y = \text{trace}\begin{pmatrix} b_x & b_y \\ -a_x & -a_y \end{pmatrix}.$$

Why does that work?

Recall the equivalence of Green's theorem and the 2-d divergence theorem.

"[A] circulation integral of $\mathbf{v} = (a, b)$ along a curve is a flux integral of $\mathbf{w} = \mathbf{v}_\perp = (b, -a)$ along the same curve; and vice versa."

More precisely, the circulation density of $\mathbf{v} = (a, b)$ along a curve is the flux density of $\mathbf{v}_\perp = (b, -a)$; and vice versa.

That means that the circulation density of $\mathbf{v}$ is the

$$\text{trace}(\mathbf{Jv}_\perp) = \text{trace}\begin{pmatrix} b_x & b_y \\ -a_x & -a_y \end{pmatrix}.$$

Bringing the two results together,

$$flux\ density = \text{trace}(\mathbf{Jv}) = \text{trace}(\mathbf{Jv})_S, \text{ and}$$
$$circulation\ density = \text{trace}(\mathbf{Jv}_\perp) = \text{trace}(\mathbf{Jv}_\perp)_S.$$

Next, let's have a look at equalities observed between density functions and rates of change obtained from geometric constructions.

In the 2-d divergence theorem, flux density $a_x + b_y$ is equal to the rate of relative change in areas.

The two quantities are computed and defined differently.

The flux density is a limit of ratios of line integrals to interior areas, with no explicit reference to streamlines.

But the rate (of change) of relative change in areas is obtained by tracking vertices of rectangles along respective streamlines, with a common parameter.

Let's explore a few examples.

Example

Suppose that the coordinate system is already aligned with the vector field. In other words, with (x, y) coordinate lines, we suppose that the vector field is of the form $\mathbf{v}(x, y) = (a(x, y), 0)$.

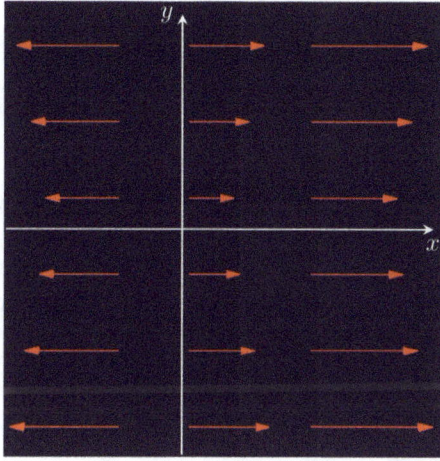

Figure 5.1   A vector field with y-component zero; $\mathbf{v} = \big(a(x, y), 0\big)$.

Let's compute both quantities, that is, both the ratio of flux to area and the ratio of relative change in areas.

We can start with a rectangle whose vertices are listed in counter-clockwise order:

$$(x, y), (x + dx, y), (x + dx, y + dy), (x, y + dy).$$

For the rectangular region $dx \times dy$ with vertices given in counterclockwise order starting with (x, y), the element of flux is given by the following computation:

$$\left[\Big(a(x, y), 0\Big) \cdot (0, -dx)\right] + \left[\Big(a(x + dx, y), 0\Big) \cdot (dy, 0)\right]$$
$$+ \left[\Big(a(x + dx, y + dy), 0\Big) \cdot (0, -dx)\right]$$
$$+ \left[\Big(a(x, y + dy), 0\Big) \cdot (-dy, 0)\right]$$

Along the upper and lower boundaries of the region, the vector field is orthogonal to the normal vectors of the boundary curves. As the last equation gives explicitly, the element of flux is, therefore,

$$[0] + \left[\Big(a(x + dx, y), 0\Big) \cdot (dy, 0)\right] + [0] + \left[\Big(a(x, y + dy), 0\Big) \cdot (-dy, 0)\right].$$

Problem 5.1.
Show that, as before, by linear approximation — ignoring high order error terms — the sum reduces to $a_x(x, y)dxdy$.

Dividing by the area $dxdy$ we get that flux density is $a_x(x, y)$ (which is $a_x + b_y$ because in these coordinates, $b \equiv 0$).

In the same example, let's look at the rate of change of the relative change in areas.

As in the general derivation of the result given earlier, we need to compute an approximation to

$$\frac{[Area]' - [Area]}{[Area]}$$

per unit time.

In the present example, motion is parallel to x axis.

Problem 5.2.
Show that the rate of change of relative area reduces to $a_x(x, y)$.

Notice that the coordinate axes are lined up with streamlines. Since motion is parallel to the x-axis, the only contribution to flux density is $a_x(x, y)$; and the rate of change of relative change of areas is $a_x(x, y)$.

Problem 5.3.
Investigate the geometry in the somewhat more general case where

$$\mathbf{Jv} = \begin{pmatrix} a_x & 0 \\ 0 & b_y \end{pmatrix} = (\mathbf{Jv})_S.$$

Problem 5.4.
Investigate the case where

$$\mathbf{Jv} = \begin{pmatrix} 0 & -f(x, y) \\ f(x, y) & 0 \end{pmatrix} = (\mathbf{Jv})_A$$

with

$$f(x, y) = \frac{b_x - a_y}{2}.$$

That is, suppose that the Jacobian represents a pure rotation. Using only elementary geometry, show that in the case of pure rotation, areas are preserved. That is, show that

$$[Area]' - [Area] = 0.$$

Hint: Remember that the components of the Jacobian of $\mathbf{v}$ are coefficients for an ordinary system of linear differential equations and initial-value problem. See Section 3.10. Observe also that $\text{trace}(\mathbf{Jv}) = 0$.

5.2 The divergence and the Jacobian

The examples in Section 5.1 draw attention to the fact that, in certain cases, it is clear from geometry that the divergence is the same as the rate of change of the relative change in areas along streamlines.

The main clue is the following:

For any two square matrices $\mathbf{A}$ and $\mathbf{B}$,

$$\text{trace}(\mathbf{A} + \mathbf{B}) = \text{trace}(\mathbf{A}) + \text{trace}(\mathbf{B}).$$

If we apply this to Jacobian matrices, we get

$$\text{div } \mathbf{v} = \text{trace}(\mathbf{Jv}) = \text{trace}\begin{pmatrix} a_x & a_y \\ b_x & b_y \end{pmatrix}$$

$$= \text{trace}\left[\begin{pmatrix} a_x & \dfrac{a_y + b_x}{2} \\ \dfrac{a_y + b_x}{2} & b_y \end{pmatrix} + \begin{pmatrix} 0 & -\dfrac{b_x - a_y}{2} \\ \dfrac{b_x - a_y}{2} & 0 \end{pmatrix} \right]$$

$$= \text{trace}((\mathbf{Jv})_S + (\mathbf{Jv})_A)$$

$$= \text{trace}((\mathbf{Jv})_S) + \text{trace}((\mathbf{Jv})_A)$$

$$= \text{trace}(\mathbf{Jv})_S) + 0$$

$$= \text{trace}((\mathbf{Jv})_S)$$

This means that, in addition to the canonical decomposition of the Jacobian into its symmetric and anti-symmetric parts, the 2-d divergence along streamlines is independent of the 2-d scalar curl.

Notice, however, that the decomposition into symmetric and anti-symmetric parts includes a_y and b_x in both summands:

$$\mathbf{Jv} = \begin{pmatrix} a_x & a_y \\ b_x & b_y \end{pmatrix}$$

$$= \begin{pmatrix} a_x & \dfrac{a_y + b_x}{2} \\ \dfrac{a_y + b_x}{2} & b_y \end{pmatrix} + \begin{pmatrix} 0 & -\dfrac{b_x - a_y}{2} \\ \dfrac{b_x - a_y}{2} & 0 \end{pmatrix}$$

Note also that the structures are linear.

Let M_2 stand for the linear vector space of real 2×2 matrices.

While the representation is not unique, every matrix in M_2 is the Jacobian of some vector field $\mathbf{v}$.

So, define the transformations $S, A : M_2 \longrightarrow M_2$ by

$$S(\mathbf{Jv}) = (\mathbf{Jv})_S = \begin{pmatrix} a_x & \dfrac{a_y + b_x}{2} \\ \dfrac{a_y + b_x}{2} & b_y \end{pmatrix},$$

and

$$A(\mathbf{Jv}) = (\mathbf{Jv})_A = \begin{pmatrix} 0 & -\dfrac{b_x - a_y}{2} \\ \dfrac{b_x - a_y}{2} & 0 \end{pmatrix}.$$

We already have that $M_2 = rgS \oplus rgA$. (See Section 3.10.)
Explicitly,

$$rgS = \left\{ \mathbf{M} = \begin{pmatrix} e & f \\ f & g \end{pmatrix} \middle| e, f, g \in \mathbb{R} \right\}$$

and

$$rgA = \left\{ \mathbf{N} = \begin{pmatrix} 0 & -h \\ h & 0 \end{pmatrix} \middle| h \in \mathbb{R} \right\}.$$

Observe also that $(A \circ S) = 0$ and $(S \circ A) = 0$.

The divergence of a vector field $\mathbf{v}$ is then given by the following composition of linear maps:

$$\operatorname{div} \mathbf{v} = \operatorname{trace}\Big(S(\mathbf{Jv}) \Big).$$

Using the abbreviation 'ker' to stand for the kernel (or null space) of a linear map, observe that for

$$(\text{trace} \circ S) : M_2 \longrightarrow \mathbb{R}$$

we have

$$\dim \ker(\text{trace} \circ S) = 3,$$

and

$$\dim rg(\text{trace} \circ S) = 1.$$

Furthermore, $rg(A) \subset \ker(\text{trace} \circ S)$ is a one-dimensional subspace of the 3-dimensional kernel of the linear map $(\text{trace} \circ S)$.

Note also that

$$[\ker(\text{trace} \circ S)] \cap A \neq 0^{35}$$

and so a more explicit identification of $\ker(\text{trace} \circ S)$ is:

$$\ker(\text{trace} \circ S) = \Big\{ \mathbf{M} = \mathbf{M}^t \Big| \text{trace}(\mathbf{M}) = 0 \Big\} \oplus A.$$

What does the analysis of linear structure mean for a fluid with a 2-d velocity field $\mathbf{v}$?

If $\text{div}\,\mathbf{v} = 0$ then, as defined in Section 4.3, the fluid is *incompressible*.

From the linear structure described above, an incompressible fluid can have a velocity field that is partly rotational.

If $\text{div}\,\mathbf{v} = 0$, then eigenvalues of the symmetric part $(\mathbf{Jv})_S$ are either both zero, or are equal but of opposite sign.

If they are both zero, then there is neither contraction nor dilation relative to the principle coordinates.

If they are non-zero but equal and of opposite sign, then there is contraction relative to one of the principle axes matched by a dilation relative to the other principle axis. See Figure 3.22.

5.3   The curl and the Jacobian

Let's now sort why the (scalar) curl, or circulation density, is a "signed off-diagonal trace" of $(\mathbf{Jv})_A$.

---

[35]For example trace $\begin{pmatrix} 2 & 5 \\ 5 & -2 \end{pmatrix}$.

Problem 5.5.

By going back to first principles, investigate 2-d (scalar) **curl** and geometry in the case where the velocity field is of the form $\mathbf{v} = (a(x, y), 0)$.

Let's do it together.

We are inquiring about **curl**. There is, however, a correspondence between curl and *divergence*, namely, $\operatorname{curl} \mathbf{v} = \operatorname{div} \mathbf{v}_\perp$. Where does that lead?

Given the vector field $\mathbf{v} = (a, b)$, then $\mathbf{v}_\perp = (b, -a)$ also is a vector field with its own Jacobian, that is,

$$(\mathbf{Jv}_\perp) = \begin{pmatrix} b_x & b_y \\ -a_x & -a_y \end{pmatrix}.$$

The Jacobian of $\mathbf{v}_\perp$ has its own symmetric and anti-symmetric parts:

$$(\mathbf{Jv}_\perp)_S = \begin{pmatrix} b_x & \dfrac{b_y - a_x}{2} \\ \dfrac{b_y - a_x}{2} & -a_y \end{pmatrix}, \quad (\mathbf{Jv}_\perp)_A = \begin{pmatrix} 0 & -\dfrac{a_x + b_y}{2} \\ -\dfrac{a_x + b_y}{2} & 0 \end{pmatrix},$$

and

$$(\mathbf{Jv}_\perp) = (\mathbf{Jv}_\perp)_S + (\mathbf{Jv}_\perp)_A.$$

It follows that

$$
\begin{aligned}
\operatorname{div} \mathbf{v}_\perp &= \operatorname{trace}(\mathbf{Jv}_\perp) && \text{(inspection of } \mathbf{v}_\perp) \\
&= \operatorname{trace}\left((\mathbf{Jv}_\perp)_S + (\mathbf{Jv}_\perp)_A\right) && \text{(decomposition of } (\mathbf{Jv}_\perp)) \\
&= \operatorname{trace}\left((\mathbf{Jv}_\perp)_S\right) + \operatorname{trace}\left((\mathbf{Jv}_\perp)_A\right) && \text{(linearity of the trace function)} \\
&= (\mathbf{Jv}_\perp)_S + 0 && ((\mathbf{Jv}_\perp)_A \text{ is anti-symmetric)} \\
&= b_x - a_y && \text{(inspection of } (\mathbf{Jv}_\perp)_S) \\
&= \operatorname{curl} \mathbf{v} && \text{(definition)}
\end{aligned}
$$

The computation also reveals that the 2-d scalar curl is unaffected by 2-d divergence along streamlines. Again, analysis of linear structures gives additional results.

Observe that

$$(\mathbf{Jv}_\perp) = \begin{pmatrix} b_x & b_y \\ -a_x & -a_y \end{pmatrix}$$

is obtained from

$$(\mathbf{Jv}) = \begin{pmatrix} a_x & a_y \\ b_x & b_y \end{pmatrix}$$

by switching two rows and multiplying one row by (-1).

Hence $(\mathbf{Jv}_\perp) = (\mathbf{QP})(\mathbf{Jv})$, where

$$\mathbf{P} = \begin{pmatrix} 0 & 1 \\ 1 & 0 \end{pmatrix} \text{ and } \mathbf{Q} = \begin{pmatrix} 1 & 0 \\ 0 & -1 \end{pmatrix}.$$

Or, $(\mathbf{Jv}_\perp) = T(\mathbf{Jv}) = (\mathbf{QP})(\mathbf{Jv})$.

The transformation $T : M_2 \longrightarrow M_2$ defined by

$$T(\mathbf{M}) \longrightarrow (\mathbf{QP})\mathbf{M}$$

is an isomorphism of M_2:

(i) If $(\mathbf{QP})\mathbf{C}_1 = (\mathbf{QP})\mathbf{C}_2$, then

$$(\mathbf{QP})^{-1}(\mathbf{QP})\mathbf{C}_1 = (\mathbf{QP})^{-1}(\mathbf{QP})\mathbf{C}_2,$$
$$\mathbf{C}_1 = \mathbf{C}_2,$$

hence T is one-one.

(ii) Given $\mathbf{C} \in M_2$, then $T((\mathbf{QP})^{-1}\mathbf{C}) = \mathbf{C}$, hence T is onto.

Problem 5.6.

Recall that if $\operatorname{curl} \mathbf{v} = 0$, then the vector field is *irrotational*. Because $\operatorname{div} \mathbf{v}_\perp = \operatorname{trace}(\mathbf{Jv}_\perp) = \operatorname{trace}(T(\mathbf{v}))$, just as was done in Section 5.2, analysis of the kernel of appropriate linear maps provides further results.

Orthogonal trajectories of v are determined by $\mathbf{v}_\perp$.

If $\mathbf{v}$ is the velocity field for a fluid then the *orthogonal trajectories* of $\mathbf{v}$ do not represent mass-flow. What do they represent? If a common parameter is used for integral curves of $\mathbf{v}_\perp$ then, physically, an orthogonal trajectory is a front-line or leading edge of advancing fluid whose streamlines are determined by $\mathbf{v}$. In important applications, $\mathbf{v} = \nabla f$. In that case, there is additional structure. This is explored somewhat in the exercises that follow.

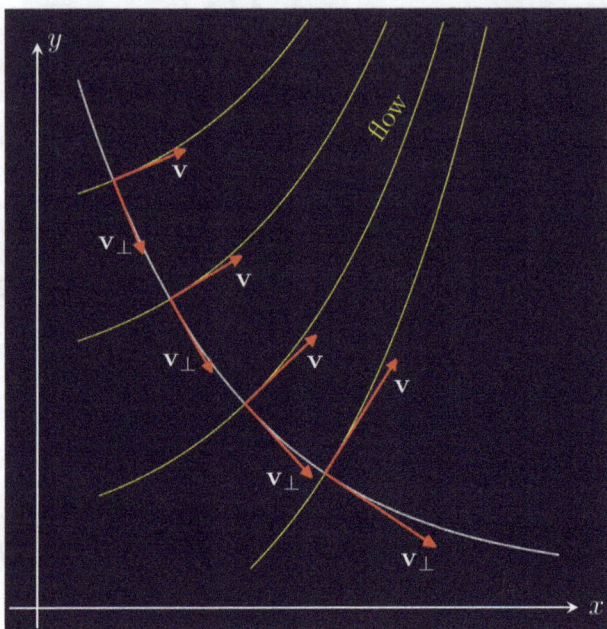

Figure 5.2   Orthogonal trajectories of $\mathbf{v} = (a, b)$ are integral curves of $\mathbf{v}_\perp = (b, -a)$.

Exercises

Exercise 5.3.1

Let $f(x, y) = x^2 - y$. Find the level curves $f(x, y) = c$, where c is a constant. Let $\mathbf{v}(x, y) = \nabla f = (2x, -1)$ and $\mathbf{v}_\perp(x, y) = (-1, -2x)$. Show that the level curves of f are the streamlines of $\mathbf{v}_\perp(x, y)$ and that the orthogonal trajectories to the level curves are the streamlines of $\mathbf{v}(x, y) = \nabla f$. Graph both sets of curves and explain.

Exercise 5.3.2

Let $\mathbf{r}(u)$ be a flow line of a gradient vector field $\mathbf{v} = \nabla f$. Show that $g(u) = f(\mathbf{r}(u))$ is an increasing function of u. Thus a particle moving along a flow line of the gradient field $\mathbf{v} = \nabla f$ will move from lower to higher values of the *potential* function f.

PART 2

FLUID MOTION IN 3 DIMENSIONS

Chapter 6

FLUX AND DIVERGENCE

6.1 Beginnings

As in the 2-d case, we start with a velocity vector field. Now, however, we look to the three-dimensional case. The velocity field is of the form $\mathbf{v} = \big(a(x,y,z), b(x,y,z), c(x,y,z)\big)$ in an (x,y,z) coordinate system. We continue to assume that velocity fields are independent of time. We also continue to assume that all functions are continuously differentiable everywhere. For brevity, we write $\mathbf{v} = (a, b, c)$, with the understanding that the coordinate functions are of the form $a = a(x,y,z)$, $b = b(x,y,z)$ and $c = c(x,y,z)$.

Problem 6.1.
What are some examples?

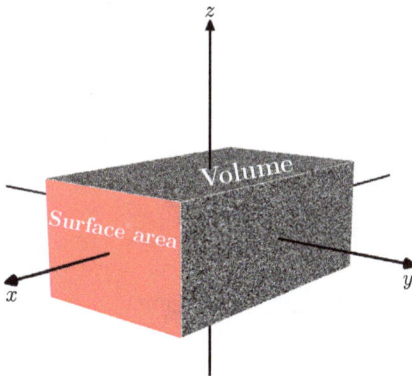

Figure 6.1   Mass per unit area is used to compute mass of volumes.

As in the 2-d case, the velocity field can be integrated to give integral curves, trajectories of "test particles." Also as in the 2-d case, initially, density functions are empirically verifiable. For instance, for a layer of known depth there is a mass-density per unit area. It is called "per unit area" with the understanding that masses are of volumes.

For a tube of known diameter, there can be a length density function. Again, mass is of increments of volume, in this case along a tube.

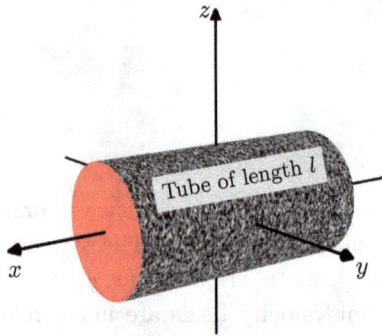

Figure 6.2   Volume has mass. Mass per unit length is used to compute mass of volumes.

With these subtleties understood, we now ask about mass-flow rates across surfaces.

Suppose that an *area density* function μ is defined on surface S in (x, y, z) space.

Mass-density is a property of a fluid. As in Part 1, we restrict to the case where that function is constant. If that constant is not unity then, by an appropriate adjustment of units, it becomes unity. Henceforth, we assume that $\mu = 1$. Under these hypotheses, this means that, numerically, momentum $\boldsymbol{\mu} = \mu(a, b, c) = (a, b, c)$. Although, *units* of momentum are *mass* × *velocity*.

Problem 6.2.
Work out details for the change of scale needed if, for instance, mass-density per unit area is gm per square cm.

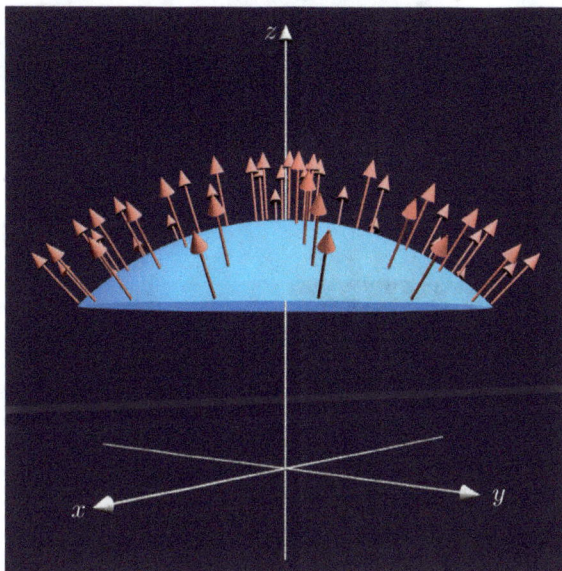

Figure 6.3 Velocity field passes through surface area.

6.2 Flux across a surface S

We now assume that surface mass-density function μ is unity, per square unit of area. Let $\mathbf{v} = \big(a(x,y,z), b(x,y,z), c(x,y,z)\big)$ be the mass-flow velocity vector at (x, y, z) and suppose that dS is the area of a small "patch" or increment of a surface S.

Question: How shall we define and compute the mass-flow rate **across** an increment of surface area? As in 2-d, a key is that, in physics, momenta add. It is that law of physics that justifies resolving a velocity vector field into components relative to a reference frame.

Let dS be the area of a small "patch" at a point (x, y, z) on the surface S. (Using linear approximation, this will be defined more carefully in sections below.) Let $\mathbf{n}$ be a *unit vector normal* to the tangent plane of S at (x, y, z). For a vector field $\mathbf{v} = (a, b, c)$, the *increment of flux* is the increment of mass flow orthogonal to S at (x, y, z). That is, $d(\text{flux}) = (\mathbf{v} \cdot \mathbf{n})dS$.

In other words, the increment of flux across that patch of surface at is (the component of $\mathbf{v} = (a, b, c)$ that crosses the surface orthogonally) $\times$ (the increment of surface area).

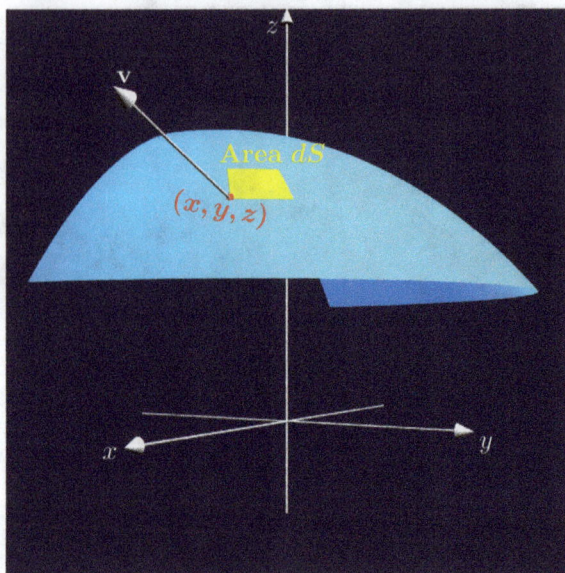

Figure 6.4   Velocity field $\mathbf{v}(x, y, z)$, patch of surface area dS.

Now that we have "increments of flux" defined, much as in the 2-d case, this allows for the following:

Definition 6.1. For $\mathbf{v} = (a, b, c)$ and suppose that $\mathbf{n}$ is a unit normal vector normal field defined along the surface S. Relative to a unit normal vector field $\mathbf{n}$, the *flux* (mass-flow rate) across the surface S is

$$\iint\limits_S \mathbf{v} \cdot \mathbf{n}\, dS.$$

The definition includes the word "relative to $\mathbf{n}$". As you may already have observed, at a point on a surface in (x, y, z) space, there are two possible unit vectors. We will need to say more about that, shortly.

As in the 2-d case, this is an average, or integral. In the 3-d case, it is an integral of mass-flow rates across the surface S in (x, y, z) space.

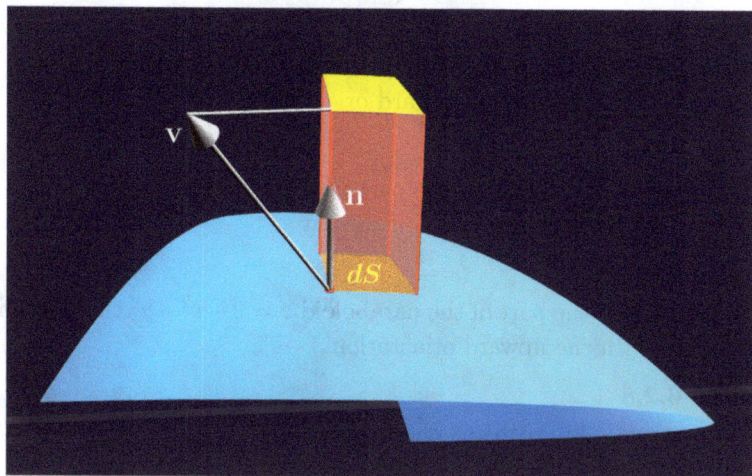

Figure 6.5 Increment of flux is increment of mass-flow orthogonal to surface, that is, $d(\text{flux}) = (\mathbf{v} \cdot \mathbf{n})dS$.

Exercises

In Exercises 6.2.1–6.2.8, find the flux for the given vector field $\mathbf{v}$ across the surface S.

Exercise 6.2.1
$\mathbf{v} = (z, 5, 2y)$, S is the part of the plane $x + y + z = 1$ in the first octant with the upward orientation

Exercise 6.2.2
$\mathbf{v} = (xy, yz, zx)$, S is the part of the paraboloid $z = 2 - x^2 - y^2$ that lies above the square $0 \le x \le 1$, $0 \le y \le 1$, and has upward orientation

Exercise 6.2.3
$\mathbf{v} = (y, xz, x)$, S is the hemisphere $y = \sqrt{4 - x^2 - z^2}$ oriented in the direction of the positive y-axis

Exercise 6.2.4
$\mathbf{v} = (x, y, z^4)$, S is the part of the cone $z = \sqrt{x^2 + z^2}$ under the plane $z = 1$ with the upward orientation

Exercise 6.2.5

$\mathbf{v} = (x, y, 2z)$, S is the part of the plane $z = 3x + 2$ that lies within the cylinder $x^2 + y^2 = 4$ with the upward orientation

Exercise 6.2.6

$\mathbf{v} = (xz, x, y)$, S is the part of the part of the sphere $x^2 + y^2 + z^2 = 81$ in the first octant with the upward orientation

Exercise 6.2.7

$\mathbf{v} = (3x, y, 5)$, S is the part of the paraboloid $z = 9 - x^2 - y^2$ that lies above the xy-plane with the upward orientation

Exercise 6.2.8

$\mathbf{v} = (xz, x, y)$, S is the hemisphere $y = \sqrt{64 - x^2 - z^2}$ oriented in the direction of the positive y-axis

6.3   Flux across a surface that bounds a volume

Example

Let $S^2 = \{(x, y, z) | x^2 + y^2 + z^2 = 1\}$, the surface of the sphere of radius 1 centered at the origin $(0, 0, 0)$. Let $\mathbf{v} = (a, b, c)$ be a velocity field.

The problem is to compute

$$\iint_{S^2} \mathbf{v} \cdot \mathbf{n} \, dS.$$

How is the definition of flux to be applied?

To compute the integral, we need to make use of parameterizations for S^2 of the form

$$\mathbf{r} : (u, v) \to \mathbf{r}(u, v) = \Big(x(u, v), y(u, v), z(u, v) \Big).$$

For the surface of the sphere, however, there is no single two-variable differentiable parameterization that covers the entire surface. Still, we can patch a few together, so long as taken together all points of the surface are accounted for.

However, to see that there is a potential problem here, let $S^2_{z>0}$ be the "north pole" hemisphere and $S^2_{x>0}$ be the hemisphere of points where $x > 0$.

A familiar parameterization for $S^2_{z>0}$ is

$$\mathbf{r}_{z>0} : (x,y) \rightarrow \mathbf{r}_{z>0}(x,y) = \left(x, y, \sqrt{1-x^2-y^2} \right).$$

In the same way, a similar parameterization for $S^2_{x>0}$ is

$$\mathbf{r}_{x>0} : (y,z) \rightarrow \mathbf{r}_{x>0}(y,z) = \left(\sqrt{1-y^2-z^2}, y, z \right).$$

Except for coordinate lines missing, $S^2_{z>0} \cap S^2_{x>0}$ is a quadrant of the surface of the sphere.

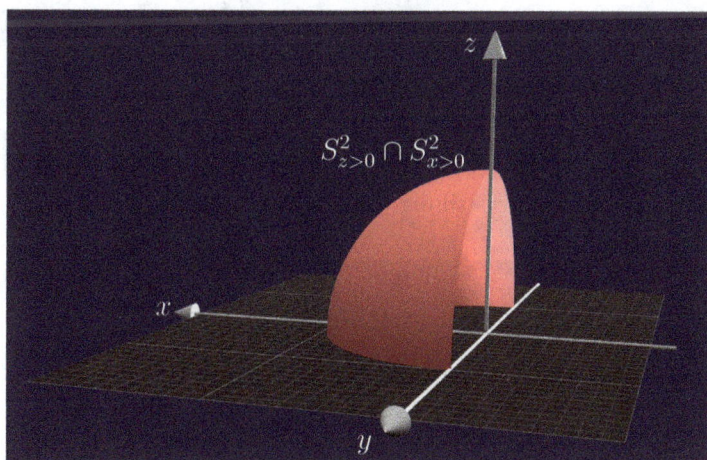

Figure 6.6   A quadrant of the surface of the unit sphere. Two parameterizations partly overlap.

Question: Which parameterization do we use for $S^2_{z>0} \cap S^2_{x>0}$?

If it doesn't matter, then the integral can be computed patch by patch, taking care not to "double count," or "triple count," etc., on regions where parameterized subsets of the surface overlap.

This raises the following more precisely formulated question:

Question: In the general case, what happens to an integral for flux when there is a change of variables?

As in the 2-d case, it is an elementary but crucial question. If you are familiar with multi-variable integration, you might correctly expect that the integral is invariant.

Problem 6.3.

Just as in the 2-d case, invariance reduces to the fact that there is a chain rule for a change variables in a surface integral. The proof is "in the books" but it is an excellent exercise to sort through the details for oneself, namely, that

$$\iint_S \mathbf{v} \cdot \mathbf{n} \, dS$$

is invariant under a change of parameterization. This means that flux is well-defined.

Preparing to continue with the example $\iint_{S^2} \mathbf{v} \cdot \mathbf{n} \, dS$ for the surface of the sphere S^2

As sample parameterizations given above already indicate, computations can be somewhat involved. So, before continuing with the example, let's pause in order to explore a similar but more elementary problem.

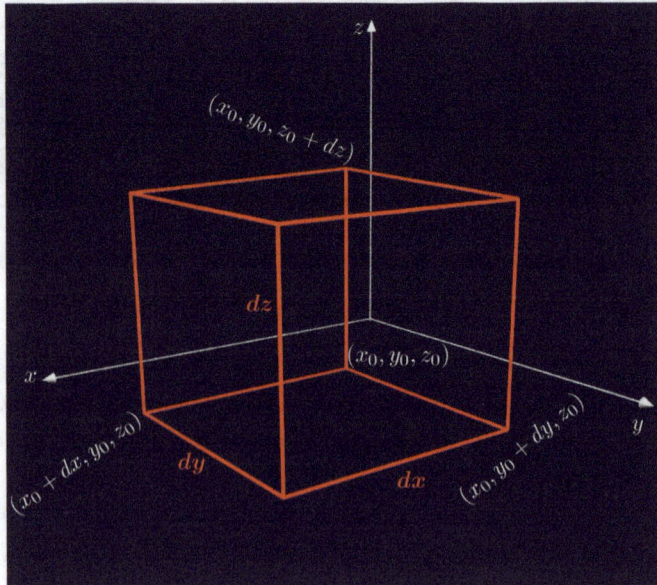

Figure 6.7   A 3-d coordinate box $dx \times dy \times dz$ based at x_0, y_0, z_0.

Let $S_{(x_0,y_0,z_0)}$ be the six-sided surface of a 3-d "box" with sides parallel to the coordinate planes, with lengths dx, dy and dz, and corner point (x_0, y_0, z_0) (see Figure 6.7).

Much as in the 2-d case, in 3-d, one way to use approximate "net flux" near a point is by using linear approximation and summing differences in flux across opposite (parallel) sides. Using that approach, net mass-flow through the 3-d box $dx \times dy \times dz$ is approximated by

$$[a(x_0 + dx, y_0, z_0) - a(x_0, y_0, z_0)]dydz$$
$$+ [b(x_0, y_0 + dy, z_0) - b(x_0, y_0, z_0)]dxdz$$
$$+ [c(x_0, y_0, z_0 + dz) - c(x_0, y_0, z_0)]dxdy$$

Problem 6.4.

Retaining only first-order terms, the "net flux" indicated is

$$[a_x dx]dydz + [b_y dy]dxdz + [c_z dz]dxdy = [a_x + b_y + c_z]dxdydz.$$

Note: We wrote "net flux" instead of '*flux*' because we were appealing to the diagram and did not use normal unit vector field **n** as is required by the definition of *flux*.

This raises a question: Can the sum of terms just obtained be expressed in terms of a normal vector field **n** defined on $S_{(x_0, y_0, z_0)}$?

Problem 6.5.

(1) The identification is made by making **n** the "outward facing unit normal" on each face. The six cases are
n $= (\pm 1, 0, 0), (0, \pm 1, 0)$ and $(0, 0, \pm 1)$.

(2) Note that **n** $= (\pm 1, 0, 0), (0, \pm 1, 0)$ and $(0, 0, \pm 1)$ as described is not differentiable. We can ignore the problematic points. What points are they? Why, for present purposes, can we ignore them?

And so we get that if $E_{(x_0,y_0,z_0)} = dx \times dy \times dz$, the coordinate box based at (x_0, y_0, z_0) , then by using linear approximation, the flux over the surface of the box is the divergence of the vector field times $dxdydz$, the volume of the box. That is,

$$\sum_{S_{x_0,y_0,z_0}} \mathbf{v} \cdot \mathbf{n} = \sum_{6 \text{ sides of box}} \mathbf{v} \cdot \mathbf{n} = [\text{div } \mathbf{v}]dxdydz. \tag{6.1}$$

Continuing with the example $\iint\limits_{S^2} \mathbf{v} \cdot \mathbf{n}\, dS$

The result just obtained is for a small coordinate box.

Can we use this to build up to a result about the surface of the sphere?

We can introduce a 3-d partition for the volume interior to the sphere, a partition consisting of boxes E_{ijk}, each of dimension $dx \times dy \times dz$ and volume $dx\,dy\,dz$. Summing over boxes E_{ijk} interior to the 3-d volume E (the unit ball) gives an approximation to the volume integral of the divergence of the vector field.

But, what do those volume sums have to do with flux along the surface S^2?

Hint: See Figure 6.8. Each box has its own unit normal vector field. Along a common face, how do unit normal vectors compare? How do surface integrals combine when obtained from a common face?

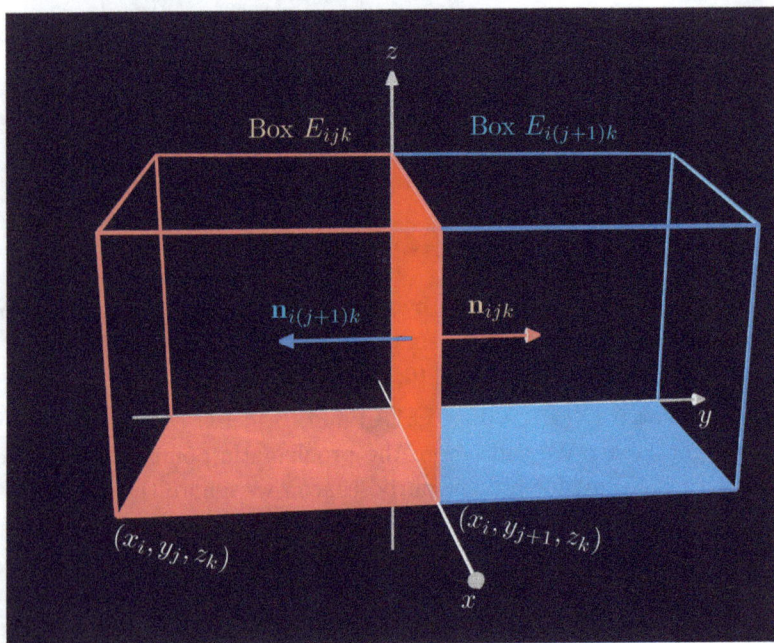

Figure 6.8   Hint for Problem 6.6.

Problem 6.6.

(1) Show that if we add equation 6.1 for two adjoining boxes in a partition, common surface integrals cancel.

(2) Repeat this result for four adjoining boxes.

(3) Extend this result to the entire triply indexed sum of boxes E_{ijk} based at (x_i, y_j, z_k) whose six-sided surfaces are labelled S_{ijk}.

The results of Problem 6.6 lead to the following two ways of evaluating the same limit:

By linear approximation, for each partition of the volume, finite sums can be written as

$$\sum_{\text{surfaces of boxes in partition}} \mathbf{v} \cdot \mathbf{n} \; = \; \sum_{\text{surfaces of boxes in partition}} (\text{div } \mathbf{v}) dx dy dz$$

with

$$\sum_{\text{surfaces of boxes in partition}} (\text{div } \mathbf{v}) dx dy dz \xrightarrow{\text{converges}} \iiint_{E=x^2+y^2+z^2 \leq 1} (\text{div } \mathbf{v}) dx dy dz.$$

From the exercise, however, cancellation of surface integrals along common faces means that the first sum also can be written as

$$\sum_{\text{surfaces of boxes in partition}} \mathbf{v} \cdot \mathbf{n} \xrightarrow{\text{converges}} \iint_{S^2} \mathbf{v} \cdot \mathbf{n} \, dS$$

And so it follows that

$$\iint_{S^2} \mathbf{v} \cdot \mathbf{n} \, dS = \iiint_{E=x^2+y^2+z^2 \leq 1} (\text{div } \mathbf{v}) dx dy dz.$$

The divergence theorem in 3-d

Various aspects of the example can be generalized.

We first need to say something about the meaning of *"positive orientation"* and *"outward facing unit normal."* In undergraduate calculus textbooks,[36] this is handled by appealing to diagrams and convention. For instance, by looking to diagrams and graphs of functions, one identifies an "outward unit normal" for an ellipsoid. But, in the textbooks, this is done by appealing to a diagram rather than a definition. Moreover, a normal vector field does

---

[36]See, e.g., James Stewart, *Calculus: Early Transcendentals*, 6th ed., (Belmont, CA: Thomson Higher Education, 2008): 1067.

not always exist. (The canonical example is the Möbius strip.) When a normal vector field does exist, it is not unique (For $\mathbf{w} \neq 0, \mathbf{n} \cdot \mathbf{w} = 0$ if and only if $(-\mathbf{n}) \cdot \mathbf{w} = 0$).

A choice of *unit normal vector field on a surface* is called an "orientation."

We won't say more on this issue here. We are drawing attention to the fact that we are not ready to rigorously define "outward unit normal." For now, we need to rely on insight into diagrams and understanding of particular cases. We hope that you are beginning to see there is a question here. But to handle the question of "orientation" would take us well beyond the context of these introductory notes.

In the same way, we won't get into attempting to rigorously define "simple solid region" in (x, y, z). Preliminary definitions can be found in Stewart's book,[37] as well as in many other Calculus III textbooks.

With these caveats, the example brings us to the divergence theorem in 3-dimesions.

Theorem 6.1. *Let* $\mathbf{v} = (a, b, c)$. *Suppose that E is a simple solid region in (x, y, z) and S is its boundary surface. This is written $S = \partial E$. Then the flux integral of $\mathbf{v}$ over the boundary $S = \partial E$ is equal to the volume integral of the divergence of $\mathbf{v}$ over the volume E.*

That is,

$$\iint\limits_{S} (\mathbf{v} \cdot \mathbf{n})\, dS = \iint\limits_{\partial E} (\mathbf{v} \cdot \mathbf{n})\, dS = \iiint\limits_{E} (div\,\mathbf{v})\, dx dy dz.$$

Problem 6.7.

Extend the method of the previous example to establish this result. In a more advanced course, rigorous limit arguments would be needed.

The solution of a technical problem that hints at the need of a higher viewpoint for surfaces and integrals

From Part 1, we have a similar integral for flux in the plane. That is, the flux of $\mathbf{w}$ across C is

$$\int\limits_{C} \mathbf{w} \cdot \mathbf{N}\, ds$$

---

[37]See note 36.

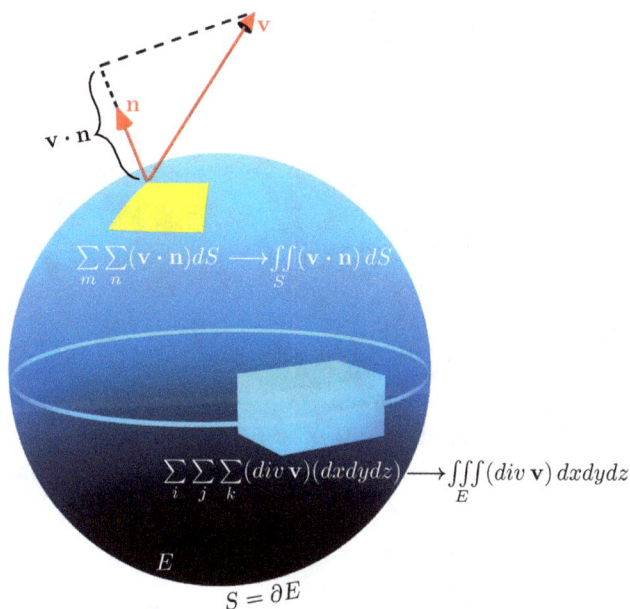

Figure 6.9   Because of linear approximation, surface integrals over boxes of the form $dx \times dy \times dz$ are volume integrals over the interiors of $dx \times dy \times dz$. Respective sums converge to the same limit, because of inner cancellations of oppositely oriented common surfaces.

where, along C, $\mathbf{N}$ is a unit vector orthogonal to the curve (tangent vector) and is a "outwardly directed." (In the present 3-d context, the symbol '$\mathbf{v}$' is already in use. So for this discussion we write $\mathbf{w} = \big(e(x, y), f(x, y)\big)$ for a 2-d vector field.)

Initially, the flux integral depends on a vector $\mathbf{N}$ that is defined by appealing to an ambient space. The question of change of variables also came up. Thanks to the chain rule, both issues were resolved:

(1) In Section 4.1, it was shown the integral can be written as a line integral without reference to a "normal vector $\mathbf{N}$"; and

(2) If two parameterizations have the same orientation along a curve, then the integral is invariant under change of parameterization. (Two parameterizations of a curve have the same orientation if at each point their tangent vectors are proportional by a strictly positive scalar function defined along the curve.)

We recall some main details of the calculation for (1), in the 2-d case:

To compute $\int\limits_C \mathbf{w} \cdot \mathbf{N} \, ds$, first obtain terms for the integrand.

These are:

$$\mathbf{w} = (e, f), \ \mathbf{N} = \frac{(y', -x')}{\| (y', -x') \|}, \ ds = \| (x', y') \| \ du, \ dx = x'du, \ dy = y'du.$$

Observe that $\| (y', -x') \| = \| (x, y) \|$.

Substituting into the integral, we get

$$\int\limits_C \mathbf{w} \cdot \mathbf{N} \, ds = \int\limits_C (e, f) \cdot (dy, -dx) = \int\limits_C -f \, dx + e \, dy.$$

In other words, the net mass-flow rate of $\mathbf{w} = (e, f)$ along C is a line integral of the vector field, without having to appeal to an additional vector defined in the ambient 2-d space.

Question: What about the 3-d case?

We have defined 3-d flux as $\iint\limits_S \mathbf{v} \cdot \mathbf{n} \, dS$.

(1) Can we write this in a way that does not require construction of a positively oriented normal vector $\mathbf{n}$?

(2) Is the integral invariant under (some) changes of parameterization?

In the 3-d case there are additional terms. But, as it turns out, we can mimic the 2-d argument.

As in the 2-d case, we begin by computing terms for the integrand. A local parameterization is needed, of form

$$\mathbf{r} : (u, v) \longrightarrow \mathbf{r}(u, v) = \Big(x(u, v), y(u, v), z(u, v) \Big).$$

For review, see a Calculus III textbook.

By definition, the term dS is an element of area of the tangent plane of S.

The tangent plane at a point has two (basis) vectors,

$$\mathbf{T}_u = (x_u, y_u, z_u) \text{ and } \mathbf{T}_v = (x_v, y_v, z_v).$$

The parallelogram determined by these two vectors has area $\|\mathbf{T}_u \times \mathbf{T}_v\|$.

The parallelogram determined by

$$\mathbf{T}_u \, du = (x_u du, y_u du, z_u du) \text{ and } \mathbf{T}_v \, dv = (x_v dv, y_v dv, z_v dv)$$

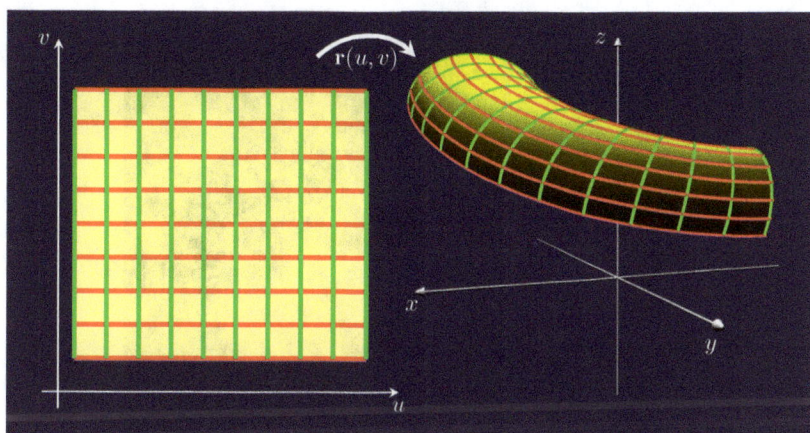

Figure 6.10 A parameterization of a 2-d surface in (x, y, z) space: $\mathbf{r} : (u, v) \to (x(u, v), y(u, v), z(u, v))$.

has area

$$dS := ||\mathbf{T}_u du \times \mathbf{T}_v dv|| = ||\mathbf{T}_u \times \mathbf{T}_v|| du dv.$$

A normal unit vector is $\dfrac{\mathbf{T}_u \times \mathbf{T}_v}{||\mathbf{T}_u \times \mathbf{T}_v||}$.

Bringing these computations together, we get

$$(\mathbf{v} \cdot \mathbf{n}) dS = \left[(a, b, c) \cdot \frac{\mathbf{T}_u \times \mathbf{T}_v}{||\mathbf{T}_u \times \mathbf{T}_v||} \right] ||\mathbf{T}_u \times \mathbf{T}_v|| du dv$$

$$= (a, b, c) \cdot (\mathbf{T}_u \times \mathbf{T}_v) du dv$$

Working this out explicitly gives

$$\mathbf{T}_u \times \mathbf{T}_v = \Big((y_u z_v - y_v z_u), (x_u z_v - x_v z_u), (x_u y_v - x_v y_u) \Big).$$

Observe that the components of $\mathbf{T}_u \times \mathbf{T}_v$ are (signed) areas obtained by projecting the parallelogram determined by $\mathbf{T}_u$ and $\mathbf{T}_v$ onto coordinate planes.

To see that, observe that the absolute value of the first component is a cross-product:

$$|y_u z_v - y_v z_u| = ||(0, y_u, z_u) \times (0, y_v, z_v)||$$

$$= \left[\begin{array}{l} \text{Area determined by the projection of the parallelogram} \\ \text{determined by } \mathbf{T}_u \text{ and } \mathbf{T}_v \text{ onto the } (y, z) \end{array} \right].$$

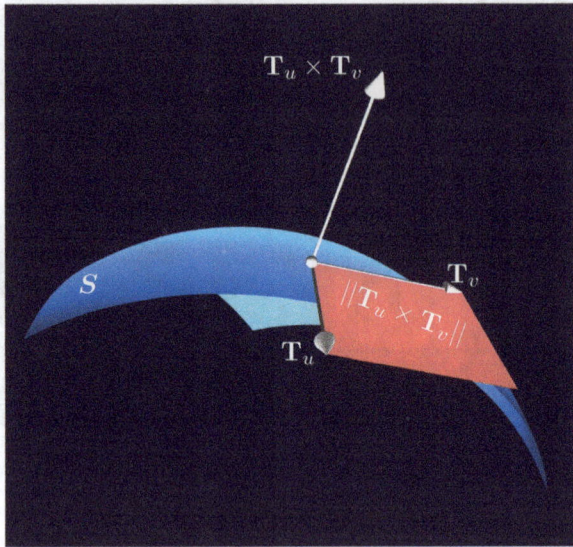

Figure 6.11   Basis vectors $\mathbf{T}_u$, $\mathbf{T}_v$ for tangent plane of S at $\mathbf{r}(u, v)$.

We also need to multiply by $dudv$.

As in the 2-d case, we can invoke the chain rule and obtain

$$[y_u z_v - y_v z_u]dudv = dydz$$

(where dy and dz are the coordinate intervals for sides of the parallelogram in (x, y, z) coordinates).

The other components are computed similarly.

That is, the second component of $\mathbf{T}_u \times \mathbf{T}_v$ is $-dxdz$ and the third component is $dxdy$.

Therefore,

$$(\mathbf{v} \cdot \mathbf{n})dS = \mathbf{v} \cdot (dydz, -dxdz, dxdy).$$

This implies that the *flux* is

$$\iint_S \mathbf{v} \cdot \mathbf{n} \, dS = \int_S \mathbf{v} \cdot (dydz, -dxdz, dxdy) = \int_S a \, dydz - b \, dxdz + c \, dxdy.$$

That is, as hoped, we obtain an expression for flux that does not explicitly involve a normal vector $\mathbf{n}$.

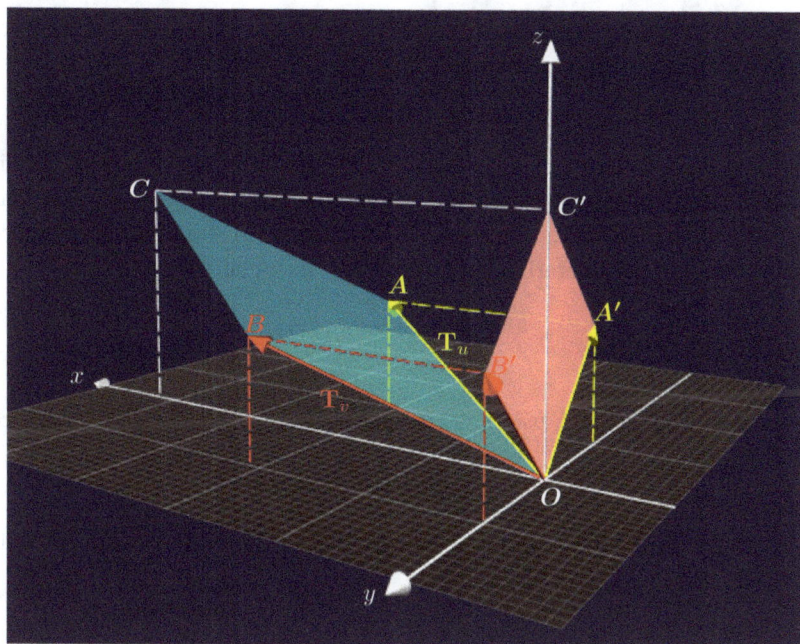

Figure 6.12 By definition, projections of $\mathbf{T}_u$ and $\mathbf{T}_v$ onto (y, z)-plane are $(0, y_u, z_u)$ and $(0, y_v, z_v)$. Area of projected parallelogram is therefore $|y_u z_v - y_v z_u|$; which is the absolute value of the first component of $\mathbf{T}_u \times \mathbf{T}_v$. Similar computations hold for the other two projections.

How is this formula used? It provides a template for evaluation of flux.

A "template"? What that means is that in each case we introduce a parameterization and for the terms $dy\,dz$, $dx\,dz$, $dx\,dy$, and use the chain rule.

Problem 6.8.
Go through the details to show that the flux is invariant under change of parameterization.

Observation: The formula

$$\text{flux} = \int_S a\,dy\,dz - b\,dx\,dz + c\,dx\,dy$$

depends on the surface S and the vector field $\mathbf{v} = (a, b, c)$ defined along S. The flux integral is, therefore, a function of "two variables," where the "variables" are surfaces and vector fields, respectively. This hints of a higher viewpoint needed to handle integrals on surfaces. For a two-parameter surface, "integrands" are of the form

$$\omega = a\,dydz - b\,dxdz + c\,dxdy.$$

These integrands are called "2-forms" and the surfaces are called "manifolds." The definition of *manifold* embraces the fact that parameterizations are not unique and that, as observed above, some surfaces (e.g., S^2) require more than one parameterization to cover an entire surface.

Exercises

In Exercises 6.3.1–6.3.4, use the Divergence Theorem to calculate $\iint\limits_{S} \mathbf{v} \cdot d\mathbf{S}$.

Exercise 6.3.1
$\mathbf{v} = (3xy^2, xe^z, z^3)$, S is the surface of the solid bounded by $y^2 + z^2 = 1$, $x = -3$, and $x = 1$.

Exercise 6.3.2
$\mathbf{v} = (x^2y, xy^2, 2xyz)$, S is the surface of the solid bounded by $x = 0$, $y = 0$, $z = 0$, and $x + 4y + z = 4$.

Exercise 6.3.3
$\mathbf{v} = (-x^4, x^3z^2, -4xy^2z)$, S is the surface of the solid bounded by $x^2 + y^2 = 1$, $z = x + 10$, and $z = 0$.

Exercise 6.3.4
$\mathbf{v} = (4x^3, 4y^3z, 3z^4)$, S is the unit sphere.

Chapter 7

STOKES' THEOREM

7.1 Stokes' Theorem: The Question

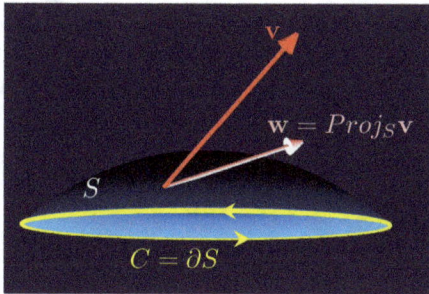

Figure 7.1 $\mathbf{w} = Proj_S\mathbf{v}$.

Suppose that S is a two-parameter surface in (x, y, z) space with boundary curve $\partial S = C$, and that $\mathbf{v} = \big(a(x, y, z), b(x, y, z), c(x, y, z)\big)$ is a vector field defined on a volume containing S.

In physical applications, questions arise about the component of $\mathbf{v}$ parallel to S.

Key point: The question is not about $\mathbf{v}$ and the surface S but is, rather, about $\mathbf{w} = Proj_S\mathbf{v}$ and the surface S.

More precisely, the inquiry is about the projection of $\mathbf{v}$ onto (the tangent plane) at each point of S.

This is similar to the problem solved by Green's Theorem, in the sense that we have a vector field $\mathbf{w}$ defined on a surface with boundary curve $\partial S = C$.

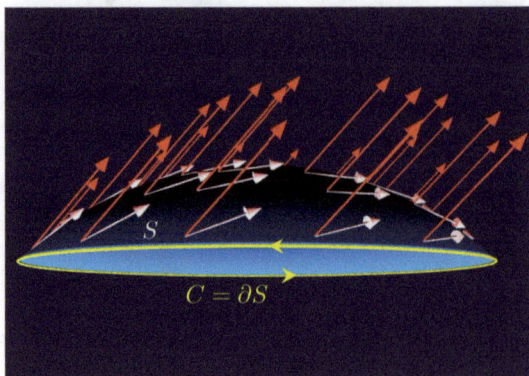

Figure 7.2   Given $\mathbf{v}(x, y, z)$ in (x, y, z)-space, $\mathbf{w} = Proj_S \mathbf{v}$ is a vector field along a 2-d surface S, with boundary curve $C = \partial S$.

Can a circulation along the boundary $\partial S = C$ be computed as an integral of a circulation density on the surface S? Is there an extension of Green's integral formula for a region in the plane to an integral formula for a region that is a surface in (x, y, z) space?

Question: In other words, is there a "Green's Theorem" for two-dimensional surfaces in (x, y, z) space?

The curl operator in the (x, y) plane cannot be applied directly to surfaces in (x, y, z) space, surfaces that, in general, can be curved and pitched at angles relative to the (x, y) plane. We will somehow need to appeal to the fact that, locally, the surface S can be parameterized by a function of the form

$$\mathbf{r} : (u, v) \to \mathbf{r}(u, v) = \Big(x(u, v), y(u, v), z(u, v)\Big). \qquad (*)$$

(1) One approach to the problem follows what we did in the development of Green's Theorem. That is, begin with the definition of circulation.

As in the 2-d case, linear approximation can be used to replace a combination of line integrals, with surface integrals. Note that since there are three components and two parameters, the computations become somewhat involved but they do not require new ideas as such.

(2) Another approach is to take advantage of (*). Then apply Green's Theorem to quantities in the (u, v)-plane.

With some guidance provided, both approaches are good homework exercises and are illuminating in different ways.

By making use of projections, the first approach calls forth techniques that eventually are needed in modern differential geometry and applications. The second approach implicitly incorporates the projection in (1) and reveals coherence of operations in vector calculus.

Below, we provide some details on both sets of computations.

7.2 Stokes' Theorem: First approach: elementary derivation of "Green's theorem" for a two-parameter surface

A first step is to compute the projection of $\mathbf{v} = (a, b, c)$ onto the tangent plane of S.[38] See Figure 7.1.

Two basis vectors tangent to the surface are:

$$T_u\Big(x(u,v), y(u,v), z(u,v)\Big) = (x_u, y_u, z_u)$$

and

$$T_v\Big(x(u,v), y(u,v), z(u,v)\Big) = (x_v, y_v, z_v).$$

We can use the cross-product to get a vector field normal to the tangent plane. That is, set

$$\mathbf{N} = \mathbf{T}_u \times \mathbf{T}_v.$$

(We use the capital letter '$\mathbf{N}$' rather than lower case '$\mathbf{n}$' because the normal vector just computed is not, in general, a unit vector as described in Section 6.2.)

Explicitly,

$$\mathbf{N} = \mathbf{T}_u \times \mathbf{T}_v = \Big((y_u z_v - y_v z_u), (x_u z_v - x_v z_u), (x_u y_v - x_v y_u)\Big).$$

The components for $\mathbf{N} = \mathbf{T}_u \times \mathbf{T}_v$ will be used below. For now, let's keep our focus on the geometry of vectors.

Using the dot product, $Proj_\mathbf{N}(\mathbf{v})$, the projection of $\mathbf{v}$ onto $\mathbf{N}$, can be computed directly.

The vector field that we need ($Proj_S(\mathbf{v})$, the projection of $\mathbf{v}$ onto the surface S) then is defined by the equation

$$\mathbf{N} = Proj_S(\mathbf{v}) + Proj_\mathbf{N}(\mathbf{v}).$$

In other words,

$$Proj_S(\mathbf{v}) = \mathbf{N} - Proj_\mathbf{N}(\mathbf{v}).$$

---

[38]You will find all of this done in Calculus III textbooks.

Using the fact that $\mathbf{v} \cdot \mathbf{N} = ||\mathbf{v}||\,||\mathbf{N}|| \cos\theta$,

$$Proj_{\mathbf{N}}(\mathbf{v}) = \left(||\mathbf{v}|| \cos\theta\right)\left(\frac{\mathbf{N}}{||\mathbf{N}||}\right)$$

$$= \left(||\mathbf{v}||\frac{\mathbf{v} \cdot \mathbf{N}}{||\mathbf{v}||\,||\mathbf{N}||}\right)\left(\frac{\mathbf{N}}{||\mathbf{N}||}\right)$$

$$= \left(\frac{\mathbf{v} \cdot \mathbf{N}}{\mathbf{N} \cdot \mathbf{N}}\right)\mathbf{N}$$

Therefore, we get that

$$Proj_S(\mathbf{v}) = \mathbf{N} - Proj_{\mathbf{N}}(\mathbf{v})$$

$$= \mathbf{N} - \left(\frac{\mathbf{v} \cdot \mathbf{N}}{\mathbf{N} \cdot \mathbf{N}}\right)\mathbf{N}$$

$$= \left(1 - \left(\frac{\mathbf{v} \cdot \mathbf{N}}{\mathbf{N} \cdot \mathbf{N}}\right)\right)\mathbf{N}$$

Let's rename the projection vector $\mathbf{w} = Proj_S(\mathbf{v})$.

The vector field $\mathbf{w}$ is defined on, and tangent to, the surface S.

The inquiry, then, is about the circulation of $\mathbf{w}$ along $C = \partial S$, the boundary of the 2-d surface $S \subset ((x, y, z)$ space).

As we did in 2-d, we can first look to a simpler case. That is, let's look to the curve on the surface determined by a positively oriented boundary of a coordinate interval.

In the present setting, a coordinate interval is $du \times dv$.

By linear approximation, we get the boundary of a parallelogram on the surface in (x, y, z) space, two of whose sides are $du\mathbf{T}_u$ and $dv\mathbf{T}_v$.

By definition of the cross-product, the element of surface area dS is $||du\mathbf{T}_u \times dv\mathbf{T}_v||$. We have an element of the surface, namely, a parallelogram tangent to the surface, with sides $du\mathbf{T}_u$ and $dv\mathbf{T}_v$. From undergraduate coordinate geometry, that planar surface element is represented by a normal vector, namely, *the surface element vector*

$$d\mathbf{S} = du\mathbf{T}_u \times dv\mathbf{T}_v.$$

As in Part 1, there are four sides, **I**, **II**, **III** and **IV**, each of which has its own initial and final points in (x, y, z) space. These are obtained by tracking counter-clockwise around the boundary of coordinate interval $du \times dv$ and consequently producing the boundary of the parallelogram in (x, y, z) space.

On sides **I**, **II**, **III** and **IV**, respectively, the contributions to circulation are:

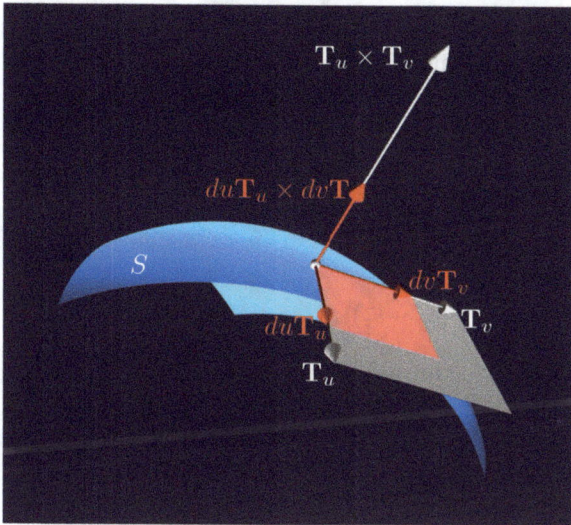

Figure 7.3   Parallelogram with sides $du\mathbf{T}_u$ and $dv\mathbf{T}_v$.

side **I**:

$$\mathbf{w}\Big(x(u,v),y(u,v),z(u,v)\Big) \cdot \Big[du\mathbf{T}_u\Big(x(u,v),y(u,v),z(u,v)\Big)\Big]$$

side **II**:

$$\mathbf{w}\Big(x(u+du,v),y(u+du,v),z(u+du,v)\Big) \cdot \Big[dv\mathbf{T}_v\Big(x(u+du,v),$$
$$y(u+du,v),z(u+,v)\Big)\Big]$$

side **III**:

$$\mathbf{w}\Big(x(u+du,v+dv),y(u+du,v+dv),z(u+du,v+dv)\Big)$$
$$\cdot\Big[(-dv)\mathbf{T}_v\Big(x(u+du,v+dv),y(u+du,v+dv),z(u+du,v+dv)\Big)\Big]$$

side **IV**:

$$\mathbf{w}\Big(x(u,v+dv),y(u,v+dv),z(u,v+dv)\Big) \cdot \Big[(-dv)\mathbf{T}_v\Big(x(u,v+dv),$$
$$y(u,v+dv),z(u,v+dv)\Big)\Big]$$

Next, collect like-terms and use linear approximation to estimate differences.

Problem 7.1.

Show that, by ignoring high order error terms, circulation around the positively oriented boundary of the coordinate interval $du \times dv$ mapped to (x, y, z) space is

$$\Big((c_y - b_z), (a_z - c_x), (b_x - a_y)\Big) \cdot \Big(du\mathbf{T}_u \times dv\mathbf{T}_v\Big).$$

The vector

$$\Big((c_y - b_z), (a_z - c_x), (b_x - a_y)\Big)$$

is called the "(3-d) vector curl."

That is,

$$\operatorname{curl}(a, b, c) = \Big((c_y - b_z), (a_z - c_x), (b_x - a_y)\Big).$$

Note: Computing the vector curl in two ways:

1. As a computational device, the components $\operatorname{curl}(a, b, c)$ can be obtained in cyclic order from the following matrix:

$$\begin{pmatrix} \partial_x & \partial_y & \partial_z & \partial_x & \partial_y & \partial_z \\ a & b & c & a & b & c \end{pmatrix}.$$

For the first component, suppress the a and ∂_x column and use the next two columns to compute $\operatorname{curl}_{(y,z)}(b, c)$.

For the second component, suppress the b and ∂_y column and use the next two columns to compute $\operatorname{curl}_{(z,x)}(c, a)$.

For the third component, suppress the c and ∂_z column and use the next two columns to compute $\operatorname{curl}_{(x,y)}(a, b)$.

The subscript notation for the pairs of variables is, we hope, clear.

This way of computing the 3-d curl vector emphasizes the fact that the components of the 3-d vector curl are 2-d scalar curls.

More precisely, the components of $\operatorname{curl}(a, b, c)$ are the scalar curls of the projections of $\mathbf{v}(a, b, c)$ onto the (y, z) plane, the (z, x) plane and the (x, y) planes, respectively.

Note also that this ordering and naming of coordinate planes matches the derivation above. To see that, write $(x\ y\ z\ x\ y\ z)$ and move from left to right in cyclic ordering. That is, first suppress x to get (y, z); then suppress y to get (z, x) ;and finally suppress z to get (x, y).

These aspects of curl $(a, b, c) = \big((c_y - b_z), (a_z - c_x), (b_x - a_y)\big)$ will be further explored in sections below.

2. Another common way to compute curl (a, b, c) is to use symbolism derived from determinants. That is,

$$\operatorname{curl}(a, b, c) = \nabla \times (a, b, c) = \begin{vmatrix} \mathbf{i} & \mathbf{j} & \mathbf{k} \\ \dfrac{\partial}{\partial x} & \dfrac{\partial}{\partial y} & \dfrac{\partial}{\partial z} \\ a & b & c \end{vmatrix}$$

where $\mathbf{i} = (1, 0, 0)$, $\mathbf{j} = (0, 1, 0)$, $\mathbf{k} = (0, 0, 1)$ and the determinant formula is applied symbolically. Observe that each row has different types of terms.

Problem 7.2 (Stokes' Theorem).

Fill in needed details. Introduce a partition of the (u, v) parameter space with coordinate intervals $du \times dv$. Observe that circulation around the boundary of S can now be computed in two ways (two different limits) and that we therefore get Stokes' theorem, namely, with all terms defined as above,

$$\int_{C = \partial S} \mathbf{v} \cdot d\mathbf{r} = \iint_{S} \operatorname{curl} \mathbf{v} \cdot d\mathbf{S}.$$

Note: Just as in the 2-d case, the key is to use linear approximation to replace differences of functions by their derivatives multiplied by appropriate coordinate interval lengths.

Problem 7.3 (A technical issue).

The computations in 1. rely on parameters (u, v). You may remember that more than one parameterization may be needed to cover a surface. Show that integrals in the formula

$$\int_{C = \partial S} \mathbf{v} \cdot d\mathbf{r} = \iint_{S} \operatorname{curl} \mathbf{v} \cdot d\mathbf{S}$$

are invariant under change of parameterization $(u, v) \to (s, t)$.

Suppose a region R_1 in the (u, v) plane and a parameterization

$$\mathbf{r}_1 : (u, v) \to \mathbf{r}_1(u, v) = \big(x_1(u, v), y_1(u, v), z_1(u, v)\big)$$

maps R_1 to S_1; and a region R_2 in the (s, t) plane and a parameterization

$$\mathbf{r}_2 : (s,t) \to \mathbf{r}_2(s,t) = \Big(x_2(s,t), y_2(s,t), z_2(s,t)\Big)$$

maps R_2 to S_2. The regions S_1 and S_2 and their boundaries $\partial S_1 = C_1$ and $\partial S_2 = C_2$ may overlap. Provide details on why the invariance of the integrals makes it possible to apply Stokes' theorem to $S_1 \cup S_2$ with boundary $\partial(S_1 \cup S_2)$.

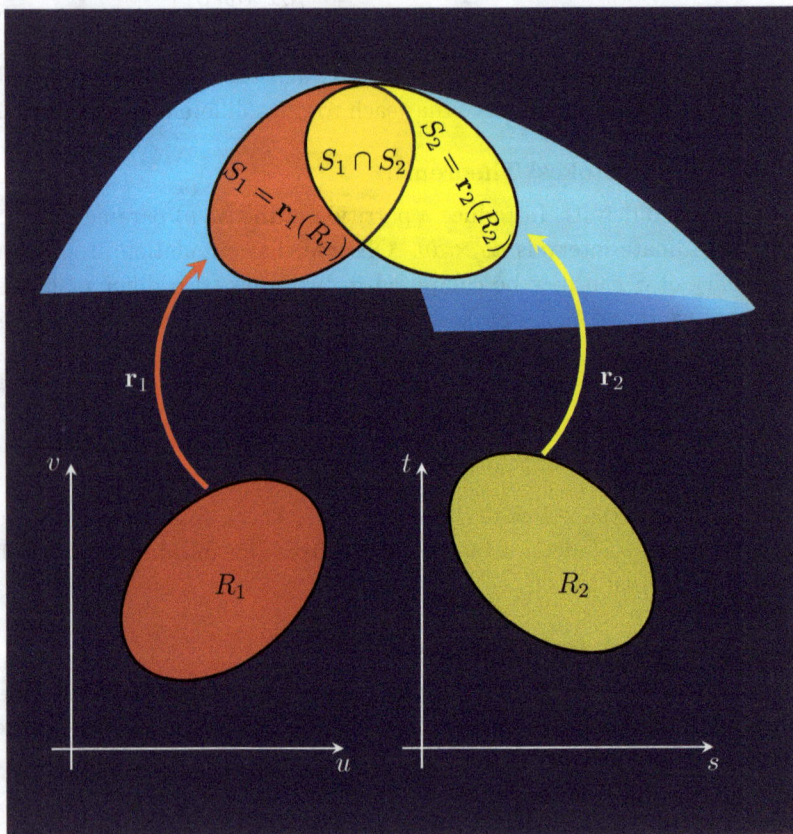

Figure 7.4   See Problem 7.3.

Without loss of generality, then, let's suppose that a single parameterization is all that is needed. Otherwise, we would patch together a finite number of parameterizations.[39] Formalizing this needs finite unions and intersections of $S_1, S_2, \cdots, S_m$.

Stokes' Theorem: Second approach: applying Green's theorem in the (u, v) parameter plane

Question: From the first approach, we can be assured that Green's theorem lifts to surfaces in (x, y, z) space. That lifted version of Green's Theorem is called Stokes' Theorem. However, can we avoid having to go back to first principles? Might we instead use what we know from the 2-dimensional case? That is, can we lift Green's theorem to a surface in 3 dimensions more directly?

The question can be put symbolically. Doing so brings focus to our search for middle terms.

$$\int_{C=\partial S} \mathbf{v} \cdot d\mathbf{r} = \text{ evaluated in } (u, v) \text{ plane;}$$

apply Green's theorem in (u, v) plane $= ? = \iint_S \text{curl}\, \mathbf{v} \cdot d\mathbf{S}.$

Let's follow our hunch. We can begin with a direct computation:

$$\int_{C=\partial S} \mathbf{v} \cdot d\mathbf{r} = \int_{\partial R} \left[\left(\mathbf{v} \cdot \mathbf{T}_u \right) du + \left(\mathbf{v} \cdot \mathbf{T}_v \right) dv \right]$$

where, as above,

$$\mathbf{T}_u = \frac{\partial \mathbf{r}}{\partial u} \text{ and } \mathbf{T}_v = \frac{\partial \mathbf{r}}{\partial v}.$$

Here, ∂R is a curve in the (u, v) plane that gets mapped to C in (x, y, z) space.

Green's theorem applies directly to the (u, v) vector field $\left(\mathbf{v} \cdot \mathbf{T}_u, \mathbf{v} \cdot \mathbf{T}_v \right)$. That is,

$$\int_{\partial R} \left[(\mathbf{v} \cdot \mathbf{T}_u) du + (\mathbf{v} \cdot \mathbf{T}_v) dv \right] = \iint_{R(u,v)} \text{curl} \left(\mathbf{v} \cdot \mathbf{T}_u, \mathbf{v} \cdot \mathbf{T}_v \right).$$

Here, $R(u, v)$ and $\partial R(u, v)$ are in the (u, v) plane.

We are attempting to make use of the parameters (u, v), so let's explore that dependence more carefully.

---

[39]Why can we assume that the number is finite? It is because the examples of surfaces given are all "compact." The definition of *compact* is provided in more advanced texts.

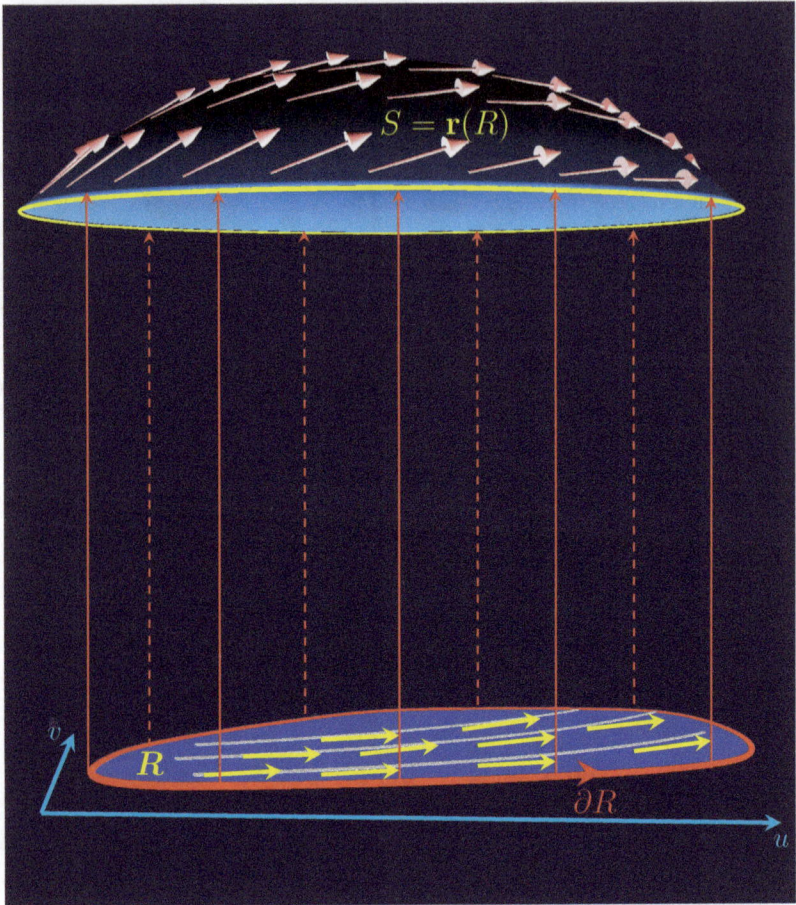

Figure 7.5   Parameterization $\mathbf{r}$ maps R in (u, v)-space to 2-d surface S in (x, y, z)-space. Express integrals defined on S in terms of parameters (u, v). Apply Green's Theorem in (u, v)-space.

In the (u, v) plane, the (scalar)

$$\text{curl}\left(\mathbf{v} \cdot \mathbf{T}_u, \mathbf{v} \cdot \mathbf{T}_v\right) = \frac{\partial \mathbf{v}}{\partial u} \cdot \frac{\partial \mathbf{r}}{\partial v} + \mathbf{v} \cdot \frac{\partial^2 \mathbf{r}}{\partial u \partial v} - \mathbf{v} \cdot \frac{\partial^2 \mathbf{r}}{\partial v \partial u} - \frac{\partial \mathbf{v}}{\partial v} \cdot \frac{\partial \mathbf{r}}{\partial u}.$$

From Calculus III, the middle terms cancel giving

$$\text{curl}\left(\mathbf{v} \cdot \mathbf{T}_u, \mathbf{v} \cdot \mathbf{T}_v\right) = \frac{\partial \mathbf{v}}{\partial u} \cdot \frac{\partial \mathbf{r}}{\partial v} - \frac{\partial \mathbf{v}}{\partial v} \cdot \frac{\partial \mathbf{r}}{\partial u}.$$

Problem 7.4.

We know what we're hunting for. So, all that we need to do is verify that the integrand

$$\left(\frac{\partial \mathbf{v}}{\partial u} \cdot \frac{\partial \mathbf{r}}{\partial v} - \frac{\partial \mathbf{v}}{\partial v} \cdot \frac{\partial \mathbf{r}}{\partial u}\right) du\, dv = (\operatorname{curl} \mathbf{v}) \cdot d\mathbf{S}\, du\, dv.$$

It then follows that

$$\iint\limits_{R(u,v)} \operatorname{curl}\left(\mathbf{v} \cdot \mathbf{T}_u, \mathbf{v} \cdot \mathbf{T}_v\right) du\, dv = \iint\limits_{R(u,v)} (\operatorname{curl} \mathbf{v}) \cdot d\mathbf{S}\, du\, dv.$$

Finally, then, we get again Stokes' theorem.[40]

The joint resolution of curl (v) and dS

Two vectors in play are

$$\operatorname{curl} \mathbf{v} = \operatorname{curl}(a, b, c)$$
$$= \Big((c_y - b_z), -(c_x - a_z), (b_x - a_y)\Big)$$

and

$$\mathbf{N} = \mathbf{T}_u \times \mathbf{T}_v$$
$$= \Big((y_u z_v - y_v z_u), -(x_u z_v - x_v z_u), (x_u y_v - x_v y_u)\Big).$$

They emerge together, by computing the *circulation density* of $\mathbf{v} = (a, b, c)$ around a surface element determined by $du\mathbf{T}_u$ and $dv\mathbf{T}_v$, a sub-parallelogram of the parallelogram determined by the two tangent vectors $\mathbf{T}_u$ and $\mathbf{T}_v$. See Figure 7.3.

By definition of the cross-product (Chapter 1), a vector that is orthogonal to the parallelogram determined by the two tangent vectors $du\mathbf{T}_u$ and $dv\mathbf{T}_v$, and whose length is equal to it, is $\mathbf{N} = \mathbf{T}_u \times \mathbf{T}_v$. Let's now compare the coordinates of $\operatorname{curl} \mathbf{v}$ and $\mathbf{N} = \mathbf{T}_u \times \mathbf{T}_v$.

The components of

$$\operatorname{curl}(a, b, c) = \big((c_y - b_z), -(c_x - a_z), (b_x - a_y)\big)$$

are the 2-d scalar curls of the projections of $\mathbf{v} = (a, b, c)$ onto the (y, z) plane, the (z, x) plane and the (x, y) plane.

---

[40] A similar derivation that uses vector product identities is: Harry F. Davis and Arthur David Snider, *Introduction to Vector Analysis*, 7th ed. (Mt. Pleasant, SC: Hawkes Learning, 2000): 298.

The components of

$$\mathbf{T}_u \times \mathbf{T}_v = \left((y_u z_v - y_v z_u), -(x_u z_v - x_v z_u), (x_u y_v - x_v y_u)\right)$$

are the signed areas of the projection of $\mathbf{T}_u \times \mathbf{T}_v$ onto the (y, z) plane, the (z, x) plane and the (x, y) plane, respectively.

We have, then, two vectors, three projections, and four areas. What are the *four* areas? They are the area A of the parallelogram determined by $\mathbf{T}_u$ and $\mathbf{T}_v$ together with $A_{(y,z)}$, $A_{(z,x)}$ and $A_{(x,z)}$, the areas of the three projections of the parallelogram of area A onto the (y, z) plane, the (z, x) plane and the (x, y) plane, respectively. See Figure 6.12.

Is there a relation between the areas A, $A_{(y,z)}$, $A_{(z,x)}$ and $A_{(x,z)}$?

If we bisect the surface area of the parallelogram determined by $\mathbf{T}_u$ and $\mathbf{T}_v$ along its diagonal, we obtain a triangle. That triangle is of area $\frac{1}{2}A$. It can be projected onto the (y, z) plane, the (z, x) plane and the (x, y) plane, providing triangles of areas $\frac{1}{2}A_{(y,z)}$, $\frac{1}{2}A_{(z,x)}$ and $\frac{1}{2}A_{(x,z)}$, respectively. An elementary computation shows that

$$\left(\frac{1}{2}A\right)^2 = \left(\frac{1}{2}A_{(y,z)}\right)^2 + \left(\frac{1}{2}A_{(z,x)}\right)^2 + \left(\frac{1}{2}A_{(x,z)}\right)^2,$$

from which it follows that

$$(A)^2 = \left(A_{(y,z)}\right)^2 + \left(A_{(z,x)}\right)^2 + \left(A_{(x,z)}\right)^2.$$

This implies that the integrand in Stokes' theorem can be written as follows:

$$\begin{aligned}
\text{curl} \cdot d\mathbf{S} = &\left(\text{curl}\, proj_{(y,z)}\mathbf{v}\right)\left(proj_{(y,z)}d\mathbf{S}\right) \\
&+ \left(\text{curl}\, proj_{(z,x)}\mathbf{v}\right)\left(proj_{(z,x)}d\mathbf{S}\right) \\
&+ \left(\text{curl}\, proj_{(x,y)}\mathbf{v}\right)\left(proj_{(x,y)}d\mathbf{S}\right).
\end{aligned}$$

where

$$\| d\mathbf{S} \| = \| proj_{(y,z)}d\mathbf{S} \|^2 + \| proj_{(z,x)}d\mathbf{S} \|^2 + \| proj_{(x,y)}d\mathbf{S} \|^2 .$$

Comment: This shows explicitly that circulation density on a 2-d surface in 3-d depends on how the 3-d vector curl $\mathbf{v}$ is pitched relative to the normal vector of the surface. This is not surprising, however, since Stokes' theorem is about the projection of $\mathbf{v}$ onto S. In particular, if $\mathbf{v}$ is parallel to $\mathbf{N} = \mathbf{T}_u \times \mathbf{T}_v$, then the projection is zero, as are the integrals.

The calculation also shows that the standard Euclidean structure of the 3-d vector $d\mathbf{S}$ is also a Euclidean structure of triples of projections of 2-d areas.

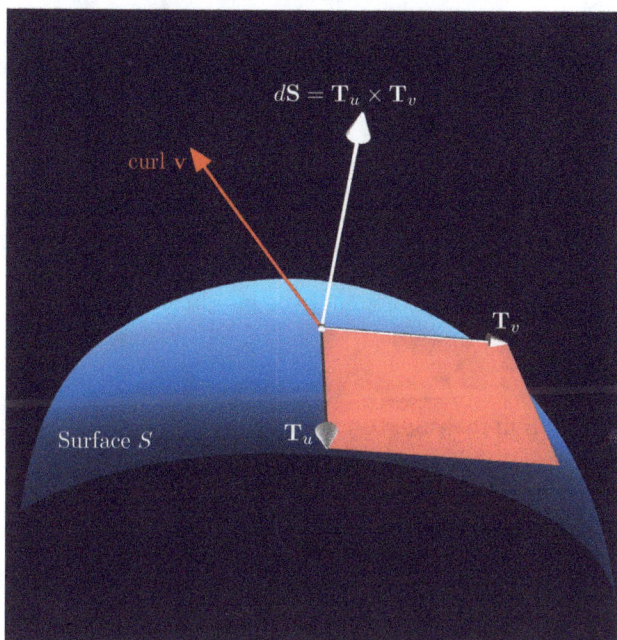

Figure 7.6   The 3-d vector curl **v** relative to the normal vector $d\mathbf{S} = \mathbf{T}_u \times \mathbf{T}_v$.

Exercises

Exercise 7.2.1
Let S_1 be the portion of the paraboloid $z = 4 - x^2 - y^2$ above the xy-plane and S_2 be the hemisphere $x^2 + y^2 + z^2 = 4$, $z \geq 0$. Suppose $\mathbf{v}(x, y, z)$ is a vector field whose components have continuous partial derivatives. Explain why

$$\iint_{S_1} \operatorname{curl} \mathbf{v} \cdot d\mathbf{S} = \iint_{S_2} \operatorname{curl} \mathbf{v} \cdot d\mathbf{S}$$

In Exercises 7.2.2–7.2.4, verify Stokes' Theorem for the given vector field **v** and surface S.

Exercise 7.2.2

$\mathbf{v} = (y, x, xz)$, S is the part of the plane $x + y + z = 1$ that lies in the first octant, oriented upward

Exercise 7.2.3

$\mathbf{v} = (y - z, x - z, y - x)$, S is the hemisphere $z = \sqrt{4 - x^2 - y^2}$, oriented upward

Exercise 7.2.4

$\mathbf{v} = (x^2 z, xy, xz^2)$, S is the portion of the paraboloid $z = 9 - x^2 - y^2$ above the xy-plane, oriented upward

Exercise 7.2.5

Use Stokes' Theorem to evaluate $\int_C \mathbf{v} \cdot d\mathbf{r}$, where $\mathbf{v} = (x^2 + 2xyz, x^2 y^2, x^2 y^3)$ and C is the intersection of $x^2 + y^2 = 1$ and $z = x - y$ oriented clockwise

Exercise 7.2.6

Let $\mathbf{v}(x, y) = \left(\dfrac{-y}{x^2 + y^2}, \dfrac{x}{x^2 + y^2}, 0 \right)$, R is the unit circle $x^2 + y^2 \leq 1$, $z = 0$. Compute both sides of the formula in Stokes' Theorem. Does this example contradict Stokes' Theorem? *Hint*: Implicit in the development of the theorem are hypotheses about domains. Also implicit is the continuity of partial derivatives of the components of $\mathbf{v}$ in xyz-space. These issues are accounted for in axiomatically rigorous presentations of the theorem.

Exercise 7.2.7

In electromagnetism, current through a surface generates a magnetic field along the curve that is boundary to the surface; and vice versa. Let $\mathbf{B}(x, y, z)$ represent the magnetic field corresponding to a steady current $\mathbf{J}(x, y, z)$. For a surface S with $\partial S = C$, the integral form of Ampère's law is:

$$\int_{C = \partial S} \mathbf{B} \cdot d\mathbf{r} = \mu_0 \iint_S \mathbf{J} \cdot d\mathbf{S}.$$

Use Stokes' Theorem to obtain the differential form of Ampère's law, that is, $\operatorname{curl} \mathbf{B} = \mu_0 \mathbf{J}$.

Chapter 8

RELATIVE CHANGE IN VOLUMES AND IN INCREMENTS

8.1 Rate of change of relative change in volume

Let $\mathbf{v}(x, y, z) = \Big(a(x, y, z), b(x, y, z), c(x, y, z) \Big)$.

In (x, y, z) space, four relatively close reference points A, B_1, B_2 and B_3 determine a small parallelepiped with sides AB_1, AB_2 and AB_3 and volume $[Volume]$.

If we (linearly) approximate the motions of all four points along their respective streamlines to A', B'_1, B'_2 and B'_3, we get a new parallelepiped with sides $A'B'_1$, $A'B'_2$ and $A'B'_3$ and volume $[Volume]'$.

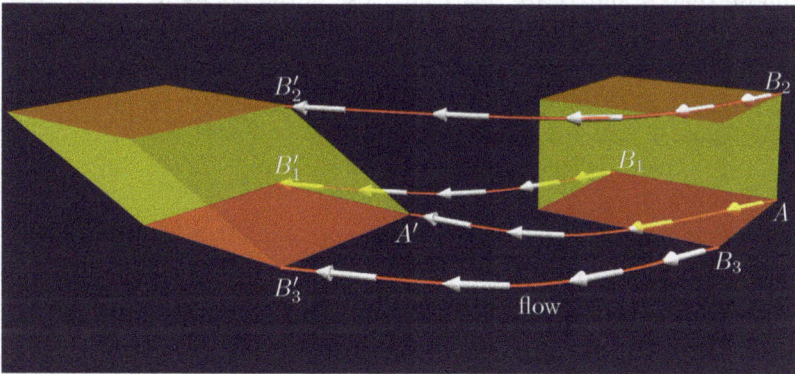

Figure 8.1   Vertices A_1, B_1, B_2, B_3 of box in (x, y, z)-space flow to vertices A'_1, B'_1, B'_2, B'_3 of parallelepiped in (x, y, z)-space.

Problem 8.1.

Using only computations from elementary geometry of vectors, and then linear approximation, show that the rate of relative change in volume is

$$a_x + b_y + c_z = \text{div } \mathbf{v} = \text{trace}\big(\mathbf{J}_{(x,y,z)}\mathbf{v}\big) = \text{trace}\big(\mathbf{J}_{(x,y,z)}\mathbf{v}\big)_S.$$

Definition 8.1. Since the rate is determined along streamlines, just as for 2-d flows treated in Part 1, a 3-d fluid flow is called *incompressible* when

$$a_x + b_y + c_z = \text{div } \mathbf{v} = \text{trace}\big(\mathbf{J}_{(x,y,z)}\mathbf{v}\big) = \text{trace}\big(\mathbf{J}_{(x,y,z)}\mathbf{v}\big)_S = 0.$$

Problem 8.2.

Explore cases where $\mathbf{Jv} = \begin{pmatrix} a_x & 0 & 0 \\ 0 & b_y & 0 \\ 0 & 0 & c_z \end{pmatrix}$. For instance, show that a rectangle lined up with the coordinate axes flows to rectangles also lined up with coordinate axes.

8.2   Rate of change of relative change of increments

As in Section 3.10, we can explore the rate of relative change of increments. Suppose that A and B are relatively close initial points in (x, y, z) space. Suppose that, up to linear approximation, A' and B' are the points to which A and B flow along their respective streamlines, in small times $dt > 0$. The problem is to compare $A'B'$ with AB. That is, the problem is to approximate the *relative strain* $A'B' - AB$.

Problem 8.3.

Show that, just as in the 2-case, linear approximation gives us that

$$A'B' = AB + \begin{pmatrix} a_x & a_y & a_z \\ b_x & b_y & b_z \\ c_x & c_y & c_z \end{pmatrix} AB\, dt = \left[\mathbf{I} + \begin{pmatrix} a_x & a_y & a_z \\ b_x & b_y & b_z \\ c_x & c_y & c_z \end{pmatrix} dt\right] AB.$$

Hence, the *relative strain* vector is given by

$$A'B' - AB = \begin{pmatrix} a_x & a_y & a_z \\ b_x & b_y & b_z \\ c_x & c_y & c_z \end{pmatrix} AB\, dt.$$

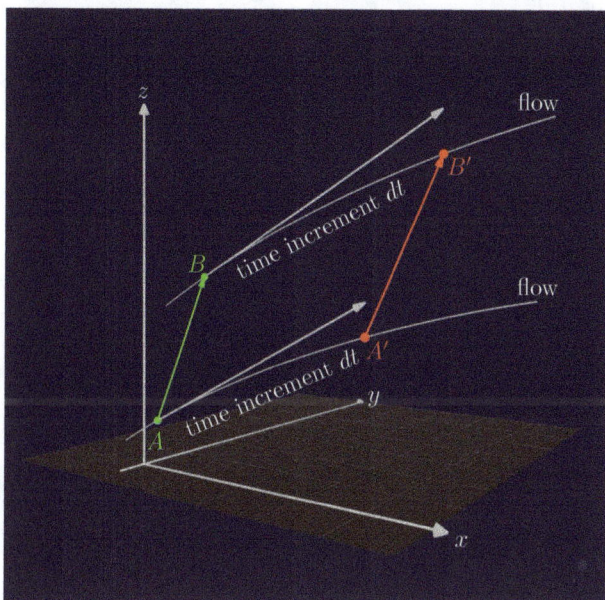

Figure 8.2   Initial points $A(x, y, z)$ and $B(x + dx, y + dy, z + dz)$ flow to A', B'. Use $\mathbf{v}(x, y, z)$ and $\mathbf{v}(x+dx, y+dy, z+dz)$ and common time increment $dt > 0$ to approximate A' and B'.

With $\mathbf{Jv}$ the Jacobian, that is

$$\mathbf{Jv} = \begin{pmatrix} a_x & a_y & a_z \\ b_x & b_y & b_z \\ c_x & c_y & c_z \end{pmatrix},$$

we get

$$A'B' - AB = (\mathbf{Jv})AB dt.$$

Problem 8.4.

As we did in the 2-d case, in a unique way, we can decompose the Jacobian into a sum of the form

$$\mathbf{Jv} = (\mathbf{Jv})_S + (\mathbf{Jv})_A,$$

where $(\mathbf{Jv})_S$ is symmetric and $(\mathbf{Jv})_A$ is anti-symmetric.
Recall that a matrix $\mathbf{M}$ is *symmetric* if and only if $\mathbf{M}^t = \mathbf{M}$; and a matrix M is *anti-symmetric* if and only if $\mathbf{M}^t = -\mathbf{M}$.
As in the 2-d case, the superscript '*transpose*' is abbreviated by '*t*'.

Show that

$$(\mathbf{Jv})_S = \frac{(\mathbf{Jv}) + (\mathbf{Jv})^t}{2} \text{ and } (\mathbf{Jv})_A = \frac{(\mathbf{Jv}) - (\mathbf{Jv})^t}{2}.$$

Observation:

$$(\mathbf{Jv})_S = \begin{pmatrix} a_x & \dfrac{b_x + a_y}{2} & \dfrac{a_z + c_x}{2} \\[2ex] \dfrac{b_x + a_y}{2} & b_y & \dfrac{c_y + b_z}{2} \\[2ex] \dfrac{c_x + a_z}{2} & \dfrac{c_y + b_z}{2} & c_z \end{pmatrix}$$

and

$$(\mathbf{Jv})_A = \begin{pmatrix} 0 & -\dfrac{b_x - a_y}{2} & -\dfrac{c_x - a_z}{2} \\[2ex] \dfrac{b_x - a_y}{2} & 0 & -\dfrac{c_y - b_z}{2} \\[2ex] \dfrac{c_x - a_z}{2} & \dfrac{c_y - b_z}{2} & 0 \end{pmatrix}$$

As in 2-d, the real symmetric matrix $(\mathbf{Jv})_S$ can be diagonalized relative to "principal coordinates." See an undergraduate linear algebra book that treats the real spectral theorem. The matrix $(\mathbf{Jv})_S$ is called the linear approximation for *pure strain* and $(\mathbf{Jv})_A$ is called the linear approximation for *pure rotation*. The reason for the name "strain" is that relative to the principle coordinates, $(\mathbf{Jv})_S$ gives the linear approximation to relative change in increments along coordinate axes. Why the name "rotation" is appropriate is brought out in Section 8.3. (Recall that Section 3.9 treats the same problem in two dimensions.)

8.3   Pure rotation $(\mathbf{Jv})_A$ represents an isometry

Notice that, except for a constant factor of $\frac{1}{2}$, the components of $(\mathbf{Jv})_A$ are the same as the components of the derived vector field

$$\text{curl}\,(a, b, c) = \big((c_y - b_z), (a_z - c_x), (b_x - a_y)\big).$$

In Section 7.2, $\big((c_y - b_z), (a_z - c_x), (b_x - a_y)\big)$ was obtained not by estimating relative changes in increments $A'B' - AB$ but by computing ratios of integrals to obtain *circulation density along a surface, per unit area*.

This leads to the following question:

Question: Why is the same vector obtained as the answer to the two different questions?

As we discussed above, there are three projections of $\mathbf{v}$ onto coordinate planes, each of which has its own scalar curl. We have already sorted out the connection between 2-d circulation and 2-d rotation. The problem here is to assemble 2-d results in the 3-d setting.

The matrix $(\mathbf{Jv})_A$ is the anti-symmetric part of the Jacobian of a velocity field, the components of which represent a system of differential equations for streamlines. For now, let's ignore the symmetric part. We will deal with that shortly.

For $(\mathbf{Jv})_A$, the system of differential equations is:

$$
\begin{pmatrix} \dfrac{dx}{dt} \\[2mm] \dfrac{dy}{dt} \\[2mm] \dfrac{dz}{dt} \end{pmatrix} = \begin{pmatrix} 0 & -\dfrac{b_x - a_y}{2} & -\dfrac{c_x - a_z}{2} \\[3mm] \dfrac{b_x - a_y}{2} & 0 & -\dfrac{c_y - b_z}{2} \\[3mm] \dfrac{c_x - a_z}{2} & \dfrac{c_y - b_z}{2} & 0 \end{pmatrix} \begin{pmatrix} x \\[2mm] y \\[2mm] z \end{pmatrix}.
$$

We treat each (two-dimensional) coordinate plane in turn.

Let $(\mathbf{Jv})_{A(yz)}$ be the restriction of $(\mathbf{Jv})_A$ to the (y, z). This is implemented by an orthogonal $\mathbf{P}^x_{(yz)}$ projection along the x axis. That is,

$$
\mathbf{P}^x_{(yz)}(\mathbf{Jv})_{A(yz)}\mathbf{P}^x_{(yz)} = \begin{pmatrix} 0 & -\dfrac{c_y - b_z}{2} \\[3mm] \dfrac{c_y - b_z}{2} & 0 \end{pmatrix}.
$$

This represents pure rotation in the (y, z) plane, with angular velocity

$$
\omega_{(y,z)} = \frac{c_y - b_z}{2}.
$$

Up to linear approximation, the projection of the flow is of the form

$$
\begin{pmatrix} y(t) \\ z(t) \end{pmatrix} = r_{(y,z)} \begin{pmatrix} \sin\left(\omega_{(y,z)}\, t\right) \\ \cos\left(\omega_{(y,z)}\, t\right) \end{pmatrix}
$$

where $r_{(y,z)}$ is the projection of the initial length $r = \| AB \| = \| (\alpha, \beta, \gamma) \|$ onto the (y, z) plane.

Similarly, we get two other coordinate plane projections of the flow:

$$\begin{pmatrix} x(t) \\ y(t) \end{pmatrix} = r_{(x,y)} \begin{pmatrix} \sin\left(\omega_{(x,y)}\, t\right) \\ \cos\left(\omega_{(x,y)}\, t\right) \end{pmatrix}$$

where $r_{(x,y)}$ is the projection of the initial length $r = \parallel AB \parallel = \parallel (\alpha, \beta, \gamma) \parallel$ onto the (x, y) plane; and

$$\begin{pmatrix} z(t) \\ x(t) \end{pmatrix} = r_{(z,x)} \begin{pmatrix} \sin\left(\omega_{(z,x)}\, t\right) \\ \cos\left(\omega_{(z,x)}\, t\right) \end{pmatrix}$$

where $r_{(z,x)}$ is the projection of the initial length $r = \parallel AB \parallel = \parallel (\alpha, \beta, \gamma) \parallel$ onto the (z, x) plane.

Each of the projections are rotational in 2-d.

What does that mean about the given flow in 3-d? Is the 3-d streamline also rotational?

The projections are known and so

$$x^2(t) + y^2(t) = r^2_{(x,y)}$$
$$y^2(t) + z^2(t) = r^2_{(y,z)}$$
$$z^2(t) + x^2(t) = r^2_{(z,x)}$$

Adding equations, we get that

$$2\left(x^2(t) + y^2(t) + z^2(t)\right) = r^2_{(x,y)} + r^2_{(y,z)} + r^2_{(z,x)}.$$

But $r = \parallel AB \parallel = \parallel (\alpha, \beta, \gamma) \parallel$. Computing the right-hand side explicitly, we get

$$r^2_{(x,y)} + r^2_{(y,z)} + r^2_{(z,x)} = 2(\alpha^2 + \beta^2 + \gamma^2) = 2r^2.$$

Therefore, in the linear approximation,

$$x^2(t) + y^2(t) + z^2(t) = r^2 = \text{ constant } > 0.$$

It follows that the flow for $(\mathbf{Jv})_A$ is an isometry in 3-d.

Problem 8.5.
The vector field $\operatorname{curl} \mathbf{v}$ is independent of the relative strain.

(i) As noted at the beginning of this section, except for a constant factor of $\frac{1}{2}$, the components of $(\mathbf{Jv})_A$ are the same as the components of the derived vector field

$$\operatorname{curl}(a, b, c) = \Big((c_y - b_z), (a_z - c_x), (b_x - a_y)\Big).$$

(ii) With $A(\mathbf{M}) = \dfrac{\mathbf{M} - \mathbf{M}^t}{2}$ and $S(\mathbf{M}) = \dfrac{\mathbf{M} + \mathbf{M}^t}{2}$, not only is $M_2 = rgS \oplus rgA$, but $A \circ S = 0$, $S \circ A = 0$, $S \circ S = S$ and $A \circ A = A$.

(iii) The $Ker\, A = rgS$ and the $ker\, S = rgA$.

(iv) There is a linear isomorphism $A(\mathbf{Jv}) \to \operatorname{curl} \mathbf{v}$.

(v) Show that the 3-d vector curl is independent of $S(\mathbf{Jv})$.

8.4 Flux density and relative change in volume

Problem 8.6.
We have two ways of defining the divergence of a vector field in 3-d: 1. as a flux density; and 2. as rate of relative change in volume.

(i) Explore cases where the Jacobian is of the form $\mathbf{Jv} = \begin{pmatrix} a_x & 0 & 0 \\ 0 & b_y & 0 \\ 0 & 0 & c_z \end{pmatrix}$.

(ii) In the general case, $\operatorname{trace}\Big(S(\mathbf{Jv})\Big) = \operatorname{div} \mathbf{v}$.

(iii) Show that the divergence is sum of the eigenvalues of $S\Big(\mathbf{J}_{(x,y,z)}\mathbf{v}\Big)$. Why does this make sense for a fluid flow? *Hint:* You will need to invoke a theorem from linear algebra.

(iv) For the general case, use the properties of the trace function to show that the 3-d divergence is unaffected by the symmetric part of the Jacobian.

Problem 8.7.
By direct computation of derivatives, show that for any vector field $\mathbf{w}$, we get $\operatorname{div}\Big(\operatorname{curl}(\mathbf{w})\Big) = 0$. Hence, if $\mathbf{v}$ is of the form $\mathbf{v} = \operatorname{curl}(\mathbf{w})$ for some vector field $\mathbf{w}$, then the flow $\mathbf{v}$ is incompressible.

Problem 8.8.

Suppose that $\mathbf{v}$ is the velocity field for a mass-flow in (x, y, z) space. For a vector field in 2-d, we identify "orthogonal trajectories," that is, "integral curves" that are everywhere orthogonal to the original vector field. (See the last paragraph of Section 5.3.) In 3-d, there is a similar construction. In 2-d, by using a common parameter, the orthogonal trajectories are moving front-lines in a mass-flow. In 3-d, we have a moving front-surface. We need the equation for the surface that is everywhere orthogonal to the mass flow given by $\mathbf{v}$. In other words, the defining differential equation for such a surface is of the form $\mathbf{v} \cdot \mathbf{f} = 0$. At each point (x, y, z), there is a two-dimensional vector space that solves the equation. The problem then is to inquire into the possibility of integrating the system to produce surfaces that are everywhere orthogonal to $\mathbf{v}$.

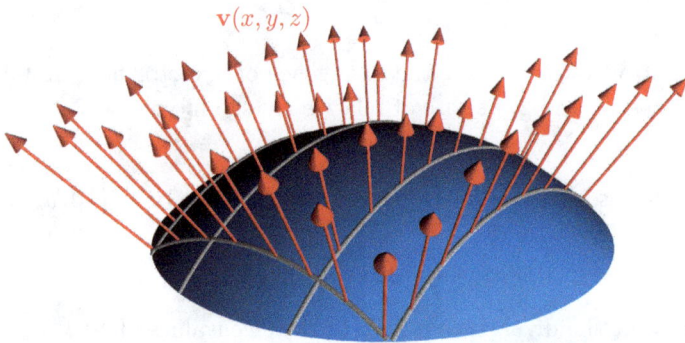

Figure 8.3   Initial points $A(x, y, z)$ and $B(x + dx, y + dy, z + dz)$ flow to A', B'. Use $\mathbf{v}(x, y, z)$ and $\mathbf{v}(x+dx, y+dy, z+dz)$ and common time increment $dt > 0$ to approximate A' and B'.

The need to study such systems of (differential) equations arises naturally in the present context. But theorems needed go well beyond basic insights in vector calculus.

PART 3

SUPPLEMENT: MATHEMATICAL UNDERSTANDING

Introduction to the Supplement

The main text was on vector calculus. Our intention is that this Supplement will be of wider application. The purpose is to help teachers and students make beginnings in describing (instances of your own) mathematical development. As will become evident, this will be useful for all levels of pedagogy.

Part of the novelty of our approach here is that, through examples and directed questions, we invite you to *grow in understanding your own mathematical understanding and that way provide yourself with essentials for growing as a teacher.*

The meaning of the italicized statement will emerge and grow by doing exercises such as those provided in Part A. What we are referring to will be a personal achievement. For the teacher, however, it is rarely an altogether private achievement. We say that because, of course, as teachers, what we think about mathematical development (which includes one's own mathematical development) factors into what we attempt to promote in our students.

Thought on teaching and learning mathematics goes back to antiquity. For instance, although in a different context and not called "mathematics education," there were Plato's reflections about the slave-boy who is helped to solve the problem of doubling the area of a square (*Meno*). However, the discipline now called "mathematics education" is mainly a 20th century development.[41] Good work has been done. For, instance, advances have

---

[41] Alan H. Schoenfeld, "Research in Mathematics Education," *Review of Research in Mathematics Education* 40, issue 1 (2016): 497-528. Philip s. Jones, ed., *A History of Mathematics Education in the United States and Canada*, Thirty-Second Yearbook. National Council of Teachers of Mathematics. Reston, VA: National Council of Teachers of Mathematics, 1970 (second printing 2002).

been made in "experiential learning" and "engaged learning." However, looking at the entire discipline, results have been mixed and opinion varies regarding the extent to which methods have been effective, or not.[42] In Part C, we provide a few comments regarding these issues.

What we would like to do first, however (in Part A), is invite you to a few relatively novel[43] exercises in mathematics. There is no "conceptual model" or "representational system." Nor do we draw on or appeal to student test scores, classroom observations, student task orientation, student cooperation, student behavior or other evidence that mathematical learning might or might not be happening for *someone else*. It's not that such results will not eventually contribute to progress in understanding mathematical development in history. The focus that we invite here is more elementary. We invite you to "go to source." The mathematics will be familiar. Part of what is new is that you are invited to a precise "puzzling about your own puzzling in mathematics." The focus, then, is *you*, and *me*, or rather, at least initially, *you-about-you* and *me-about-me*. The invitation is to make initial progress in being able to *advert* to,[44] focus on, inquire about and discern orderings of distinct events in our own inquiry and understanding, in instances, when we are doing mathematics.[45]

---

[42]Dawn Leslie and Heather Mendick, eds., *Debates in Mathematics Education*. Abingdon, Oxon: Routledge, 2014.

[43]The approach is not original but will be new for the science of mathematics pedagogy. An advanced source is: Bernard Lonergan, *Insight: A Study of Human Understanding*, eds. Frederick E. Crowe and Robert M. Doran, vol. 3 in The Collected Works of Bernard Lonergan. Toronto: University of Toronto Press, 1992. Introductory level works are: (1) John Benton and Terrance Quinn, *Journeyism*, 2018, https://bentonfuturology.com/journeyism/; (2) John Benton, Alessandra Drage and Philip McShane, *Introducing Critical Thinking*. Halifax, Canada: Axial Publishing, 2005 (Reprint 2006). (This book has been translated into Spanish (Madrid: Plaza y Valdés, 2011); and (3) Philip McShane, *Wealth of Self and Wealth of Nations: Self-Axis of the Great Ascent*, Hicksville, NY: Exposition Press, 1975. (This is available for free at http://www.philipmcshane.org/wp-content/themes/philip/online_publications/books/wealth.pdf.) Compared to the present Supplement, references (1), (2) and (3) are far broader in their coverage. The Supplement focuses on mathematical development and pushes into further examples.

[44]The (intransitive) verb 'advert' is convenient. We are using it in its first meaning in the Merrian-Webster: "advert, *intransitive verb*. 1: to turn the mind or attention - used with to."

[45]Occasionally, we find it convenient to use a single expression to refer to the whole inquiry: "self-attention." This is meant in the precise sense that will begin to emerge by doing the exercises in Part A. The turn "to attend to one's own understanding" can seem strange if one has been trained to focus mainly on models.

We introduce a key diagram (Figure A.1) which could — in a sense — be said to be a "model." But it is not a "conceptual model" or "speculative framework." Think more of something analogous to what the physics community has been getting to, namely, a "best-to-date standard model" which, for the most part, was discovered in and continues to be verified (or not) *in instances, in experience*. Or again, in chemistry, one could say that the periodic table is a "model." However, the periodic table emerged from and has been established through centuries of ongoing experimental work investigating chemical dynamics. In a somewhat similar way, we think that you will begin to see that Figure A.1 also is no mere model. By turning attention to our own experience, in exercises in Part A, Figure A.1 emerges and is verified. It too is a "table of elements," that is, "elements of dynamics of knowing." Note that Part A is only a first few "experiments." We leave it to interested readers to go further, to begin exploring the significance of Figure A.1 in other instances in your own mathematical development (and more).[46]

What all of this may have to do with improving teaching in mathematics will be touched on briefly in part B. Part C draws attention to a few anomalies in contemporary mathematics education — in both content and method. A challenge for the education community will be to take advantage of good work that has been done but to also make progress in getting a handle on the various anomalies and misdirects. To do so effectively will not be easy. It will need global collaboration. For this Supplement, it will be enough to tease a few key issues "into view."

---

[46]See references in note 42.

PART A

Mathematical Understanding

A.1 A diagram

Much in the way an introductory level chemistry text may include a simplified periodic table in the front cover, we begin by providing a simplified version[47] of a key diagram, a "table of elements of knowing." The diagram will be developed as we go.

---

[47] A more nuanced version is available in Bernard J. F. Lonergan, "Appendix A, Two Diagrams," *in Phenomenology and Logic: The Boston College Lectures on Mathematical Logic and Existentialism*, eds. Frederick E. Crowe, Robert M. Doran, vol. 18 in The Collected Works of Bernard Lonergan, Toronto: University of Toronto Press, 2001, 319–323. A diagram for the "Dynamics of Doing" is the second diagram in Lonergan, "Appendix A, Two Diagrams." Evidently, our "dynamics of doing" are similar to our dynamics of knowing. For decision, however, inquiry is in a different mode. Our dynamics of doing subsumes our dynamics of knowing. The interweaving of the two modes of inquiry are brought out in student exercises in John Benton and Terrance Quinn, "The Dynamics of Doing," *Journeyism* 16 https://bentonfuturology.com/journeyism16/ and *Journeyism 17*, https://bentonfuturology.com/journeyism17/. The dynamics of doing are not an immediate focus of this Supplement. However, it is worth noting that progress in adverting to and describing dynamics of doing also is needed. Among other things, such progress will help resolve numerous contemporary issues in mathematics education where, so far, dynamics of the two modes are not adverted to and consequently, not yet adequately distinguished. For instance, it is sometimes said that a teacher's job is to help a student "decide" which concept or solution to accept. The plausibility of such notions emerges from the mistaken model that mathematical understanding is a matter of connecting concepts. The fact that that view is mistaken is revealed — becomes (self-) evident — by doing the exercises given throughout Part A. For context in mathematics education, see Part C.

Wonder   Is it so? $\longrightarrow$!Reflective insight $\longrightarrow$ Judgement ("Yes/No/Maybe")

$\uparrow$

Wonder   What is it?$\longrightarrow$!Direct insight $\longrightarrow$ Inner formulation

$\uparrow$

Wonder   Sense

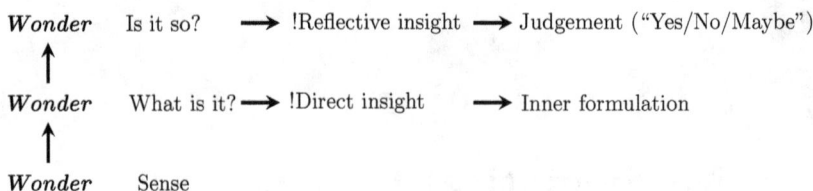

Figure A.1   Dynamics of knowing

A.2   Some puzzles

A.2.1   *A first puzzle*

A first puzzle is given by the following:

| 1 | 4 | 7 | | 11 | 14 | 17 | | 41 | 44 ... |

2 3   5 6   8 9 10   12 13   15 16   18 19 20 ... 40   42 43

An ellipsis "$\cdots$" means that the pattern is to be continued. An insight is needed. Let's concentrate on the second ellipsis, the "$\cdots$" that follows '44.'

What?-ing and direct insight

Before getting too far into solving the puzzle, let's pause in a first effort to notice something about what we are doing.

Are you wondering about "the array"? 'What is it?', where the 'it' is the array that you have in sight.

If Yes, then you are "What?-ing", you are in what we could call a "What is it? *inquiry-poise*." Why do I say "*inquiry-poise*" and not "inquiry"? Of course, both are correct. It is for emphasis only that we are adjusting familiar vocabulary somewhat. One wonders. In other words, 'what?-ing' is a transitive verb, something that we do and what?-ing is a holistic poise.

Again, we are not suggesting that we necessarily utter words such as 'What is it?'. In wondering about a seen sequence of symbols we are focused on symbols. Symbols are in and of our senses, are in and of our *sense-ability*. The temporary neologism, then, is to help draw your attention, your self-attention, to the fact that inquiry is inquiry of a whole person and where, in the present example, focus is on something in-and-of one's sense-ability.

Writing down what we have just described, we get part of Figure A.1:

Wonder   What is it?

↑

Wonder   Sense (layout of symbols, insight)

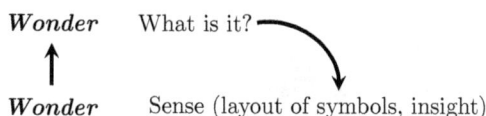

Figure A.2   Wonder about data of sense.

Perhaps you have already discovered a (possible) solution and can now continue the sequence.

If so, something happened. There was a change "in" you. A name for that "all-of-you-event" is *insight*.

Let's call this a *direct insight*.

Why do we need the adjective 'direct'? Partly, it is to distinguish this insight from what turns out to be a second kind of insight that can emerge in a follow-up inquiry-poise discussed below.[48]

For this sequence-puzzle, if you've had a direct insight, then you've gotten hold of something, a possibility.

From that (act of) insight, automatically as it were — but with no implication here of machinery or technology — there is a "procession" in you.

Why do we say "procession"? Focusing on an image in one's sense-ability, being "lit up" by insight "into" an image in one's sense-ability, not only does a light "go on" in us in that concrete image-focus, but light immediately also "goes on in us." There is in us, as it were, "light from light" in the sense that following insight, a solution "emerges in us." There is the famous story of Archimedes, his calling out "Eureka!" As the Greek verb says, when we discover the solution to the sequence puzzle, it is not just an insight. Forthwith, and forth and with, there is also an "I've got *it*." We are not playing a grammar game here but are inviting your attention to an event. There is an 'it' in "I've got it." There is something that is not merely what we had in our sense-ability. There is something that emerges "in us" from our insight-into-image, something that in many respects is "transferable." For instance, following direct insight, we can turn our attention elsewhere but then later return to the sequence-puzzle. The same solution then comes to mind but without labor.

Why is it without further labor? Because we already get the point. We already "got it." It will be convenient to introduce a name. For historical

---

[48]See below: Is?-ing and reflective insight.

reasons, let's call the "something that emerges in us," the "it" of "I've got
it," "inner formulation."[49]

Note that the adjective "inner" is not to suggest that what we get hold of
is somehow "spatially inside" (versus a "spatially outside"). The adjective
is merely a metaphor. Inner formulation is a fact. It is what we have
in as much as we have, in fact, discovered a possible solution. Note that
inner formulation is to be distinguished from "formulation" in the sense of
providing a "formula." Although, evidently, in some cases, a formula can
precede insight, and also be an expression of inner formulation.

An expression for what we have just described is given by the bottom two
rows of Figure A.1:

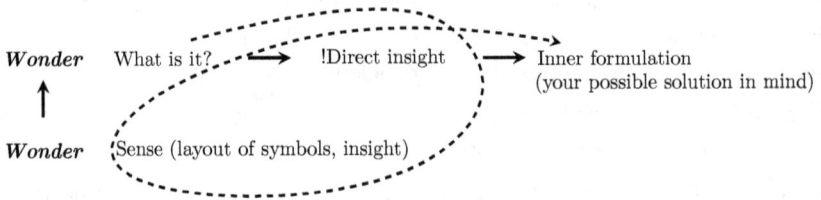

Figure A.3   Wonder about data[50] of sense, insight and inner formulation

Is?-ing and reflective insight

Does the diagram in Figure A.3 express elements that emerge in our inquiry-
poise?

Asking that question reveals that in addition to wondering 'What is it?'
we also sometimes ask a further question. That is, we also wonder 'Is *it*
so?', where the 'it' of 'Is *it* so?' is, again, our possible solution, our inner
formulation that emerged in us from our What is it? inquiry-poise.

To be sure, our Is it so? inquiry-poise does not always emerge! Whether because of haste or temperament or circumstances, sometimes we don't get to
'Is it so?' But is it not (self-)[51] evident that we do — at least sometimes —
rise to an 'Is it so?' inquiry poise? Is it not so?

---

[49]Lonergan, *Phenomenology and Logic*, Dynamics of Knowing, "Appendix A," 322. See
also McShane, *Wealth of Self*, 15.

[50]See, e.g., Benton and Quinn, "The Dynamics of Knowing: The First Four Boxes,"
https://bentonfuturology.com/journeyism10/; and McShane, Benton, Drage and Mc-
Shane, ch. 16, "What-Questions," 67-69.

[51]See note 45.

As we find in experience, then, in addition to a 'What is it?' inquiry-poise we also sometimes engage in a further inquiry-poise. We have gotten hold of something through direct insight. We might also reflect on our (potential) resolution to our What is it? inquiry-poise. To be in keeping with tradition, let's call this further inquiry *reflective inquiry*.

Are direct and reflective inquiry so different?

At a dinner party, a colleague asked a mutual friend, "What would you like to drink?" For fun, she answered, "Yes." "Is that so?" She answered, "Tea." It was funny at the time.

The purpose of our story is to help bring out that it is (self-) evident that wondering 'What is it?' and wondering 'Is it so?' are (radically[52]) different questions.

In the sequence puzzle, de facto, the question 'What is that pattern?' is not resolved by answering 'Yes.'

Nor do we resolve an 'Is it so?' question by providing a possible solution to the puzzle.

In other words, it is not merely Yes or No or Probably Yes, or Probably No. We move to an inner Yes (or inner No) *regarding a possible solution* already obtained through direct insight.

In an 'Is it so?' inquiry-poise, what is it that we are we looking for? Continuing with the sequence puzzle, one might, for instance, re-examine the pattern in one's sense-ability, to confirm whether or not one's solution holds up, that is, whether or not one can indeed account for all terms of the sequence. Or, perhaps you feel that you were distracted at the time and don't trust your solution. You might then give the whole sequence-puzzle a fresh look. And so on.

And so on? Notice that our 'Is it so?' inquiry-poise is open ended. You have discovered a possible solution. You have that solution in mind. But 'Is it so?' All that you have done so far is up for grabs. Does your solution hold up under scrutiny? Were you hasty and do you need to start over? Eventually, if one is to not simply guess, something more is needed. What is that "more" that is needed? At what point are we "good to go" (or not) with our possible solution?

In an 'Is it so?' inquiry-poise, we may have another insight. We may reach an understanding to the effect that: Yes, (probably) I have enough to go

---

[52]I am using the first meaning of the word 'radical,' referring to 'root.'

on. "We reach an understanding that (probably) we have sufficient grounds to assent to our solution." (Note that in this context the word 'probably' is an adjective, not a term from empirical probability theory. Why introduce the adverb? Again, reflection is open ended. We can always ask more questions.)

This can be as spontaneous and rapid as hearing sounds and "immediately" knowing that children are playing soccer in a field across the road. But in mathematics, modern rigor asks us to, as much as possible, spell out what precisely we understand and to identify grounds for what we understand.[53]

In the case of the sequence puzzle the pattern is what we might call a "pattern in symbolism as symbolism."[54] In attempting to resolve our 'Is it so?' inquiry, we again can advert to sense-ability. Now, though, it is not adverting to an image in a 'What is it?' inquiry-poise. Our new lean is to discern whether or not our already discovered solution accounts for all terms of the puzzle. It is a personal achievement: *one* returns to one's puzzle to sort out whether or not *one* can now, in fact, account for all terms of the puzzle.

If you are starting to become somewhat familiar with the more complete focus (that is, of not only doing mathematics but pausing to discern what you are doing when you are doing mathematics), then you may have already noticed that following reflective insight there is a second procession. The second procession is to an inner event called "judgement," an inner "*Yes*, it is so" (or "*No*, it is not so," or *Probably Yes*, or *Probably No*, where, again, in this context, 'probably' is an adverb for the quality of one's judgment). As in procession that follows direct insight, following reflective insight, procession is "automatic" as it were—it occurs. Following reflective insight, procession is to a distinct (but not separable) inner event. In some cultures, occurrence of assent (or its opposite) is obvious because it is spontaneously revealed by a familiar movement of the chin. However, becoming acquainted with people from different parts of the world soon reveals that what that movement is varies among cultures.

Until now, we have held back the solution to the sequence puzzle A.2.1. We did not want to deprive you of the pleasure of discovery.

The sequence $1, 2, 3, \cdots$ is partitioned. Some of the numbers are above a line and some are below. Why? Perhaps surprisingly, a clue is that

---

[53]That is not easy. See, e.g., note 69.
[54]If you haven't already solved the puzzle, that is a clue. See also the personal anecdote at the end of this section.

problem can be solved by school children. In case you haven't yet solved the problem, we provide the answer below. But in case you wish to keep working on it, we put the solution in a footnote to make it easier to not peek.[55]

Yes, children can solve the problem. But we are inviting you to do something more, something that children are not asked to do (at least not yet, at this time in history). We are inviting you to make beginnings in discerning aspects of what you are doing, what you have done, what you have achieved. You might notice, for instance, that in this case the key insight is "understanding differences in symbols as symbols," that is, getting hold of a pattern in one's sense-ability. If we were to write the ("natural") numbers in Roman Numerals, base 2 or (like the Babylonians) base 60, this puzzle would be destroyed.

There are, however, sequence puzzles that call for other kinds of insight.

A.2.2 *A Famous Sequence*

There is the famous sequence, $1, 1, 2, 3, 5, 8, \cdots$.[56]

If you have already gotten hold of the pattern, notice that, by contrast with the puzzle given in Section A.2.1, what is key here is not understanding what symbols happen to look like, "as symbols."[57] And so, for instance, essentially the same puzzle can be expressed in Roman numerals, base 2 and base 60.

In Roman Numerals, the puzzle can be written as: I, I, II, III, V, VIII, XIII, ...

In a base 2, the puzzle can be written as: $1, 1, 10, 11, 101, 1000, 1101, \cdots$.

---

[55] Answer to sequence puzzle A.2.1: Symbols that are written only with straight edges go above the line, while symbols that include curved edges go below the line.

[56] The Fibonacci sequence is a sequence of numbers of fertile pairs of male and female rabbits. Start with the hypotheses given by Fibonacci (about fertile male and female pairs, gestation periods and the numbers of progeny). The problem is within reach of the contemporary senior high school or undergraduate student. It is interesting and well worth doing. Discovering the solution for oneself one takes a step in a climb toward the modern theory of recursive sequences.

[57] Long before Fibonacci, the same sequence is found in ancient Sanskrit texts, written in a base 60 number system. See, e.g., Tia Ghose, "What is the Fibonacci Sequence?" *LiveScience* (October 24, 2018): https://www.livescience.com/ 37470-fibonacci-sequence.html; and Keith Devlin, *Finding Fibonacci: The Quest to Rediscover the Forgotten Mathematical Genius Who Changed the World.* Princeton, NJ: Princeton University Press, 2017.

If you have already studied some of the early history of mathematics, you will know that the sequence could also be expressed in cuneiform, using base 60.

An understanding of the sequence can also be expressed implicitly with visibly different formulas, three examples of which are:

$$x_{n+2} = x_{n+1} + x_n, \ n = 0, 1, 2, \cdots ; \tag{A.1}$$

$$f_k = f_{k-1} + f_{k-2}, \ k = 2, 3, 4, \cdots ; \text{ and} \tag{A.2}$$

$$\Re_{m+20} = \Re_{m+19} + \Re_{m+18}, \ m = -18, -17, -16, \cdots .^{58} \tag{A.3}$$

A.3 Descriptive and explanatory understanding, and judgment in mathematics

Descriptive and explanatory understanding

In example (A.1), our understanding is *descriptive*. For the sequence-puzzle A.2.1, *descriptive* refers to the fact that what we grasp is "a pattern in our sense-ability."

By contrast, in example (A.2), our understanding is *explanatory*. Rather than "a pattern in sense-ability," the key insight needed is to discover a pattern of mutually defined terms, and operations.[59]

As is evident from these two examples, both *descriptive* and *explanatory* understanding are part of and contribute to our mathematical development.

For another example, you may recall a familiar sequence from calculus that is often called the 'power rule':

$y = x, \ y' = 1; \ y = x^2, \ y' = 2x; \ y = x^3, \ y' = 3x^2;$ and so on.

---

[58]From a "higher viewpoint," we could also identify the solution as follows: "The sequence in (A.2) is the (unique) solution to a second-order homogeneous recurrence relation with constant coefficients whose (non-reduced) characteristic equation is $t^2 - t^2 - 1 = 0$, and whose first two initial-values are 1 and 1." See Section A.6.

[59]More generally, descriptive understanding is "grasping patterns in experience, as experience." And so to discern events and orderings of events in our dynamics of knowing is a beginning but it, too, is descriptive. Progress in explanatory understanding of our dynamics of knowing will be future growth for the academic community. Aspects of that future progress are already partly in evidence. And so, eventually, it will include an "integration" of human biophysics, biochemistry, human zoology and psychology, cognitional theory and more. To glimpse something of the challenge, precise densely expressed heuristics can be found in Lonergan, *Insight*, 489 (add the word 'self' to the paragraph that begins "[(Self-) S]tudy of an organism begins ...").

The symbolic pattern can be grasped without needing to understand anything at all about differentiation. In that case, what is grasped is a "symbolic pattern as a symbolic pattern," "a pattern in our sense-ability." But the mathematical meaning of these symbols is reached through explanatory understanding.

Evidently, the observation applies generally. We can understand how to use symbolism in linear algebra; differential equations; and on into the most remote realms of contemporary mathematics. In each case, descriptive understanding is an achievement and by it we understand, for instance, how to use symbols and diagrams. However, there is also the possibility of understanding what symbols and diagrams mean which, in mathematics, is explanatory understanding.[60]

Judgment

Note that in (A.1) and (A.2) we find the same core pattern expressed by Figure A.1. We begin with inquiry about an "image" in sense-ability, and "a focus inquiry." In both cases, a 'What is it?' inquiry-poise is resolved through direct insight, following from which there is a procession "in us" to "inner formulation."

At the same time, there are significant differences in what is being understood. As you might expect, grounds for judgment also turn out to be different.

In example (A.1), direct insight is of a "pattern grasped in our sense-ability." Is our solution correct? Is it so? By adverting again to patterns in our sense-ability, we find and discover sufficient grounds (or not) for assenting to our possible solution. There is then a procession to "Yes, it is so," or perhaps, "No it is not so."

In example (A.2), however, what we discover is a pattern of mutually defined terms, and operations that, as it happens, can be both presented and expressed in many ways.[61] In that case, sufficient grounds for assenting to our possible solution are not "patterns in sense-ability" but whether or not one can, indeed, successfully continue the sequence by appealing

---

[60]For a discussion in elementary contexts, see Terrance J. Quinn, "On Two Types of Learning (in mathematics) and Implications for Teaching," *On Learning Problems in Mathematics, Research Council on Mathematics Learning*, Mathematical Association of America, FOCUS, Fall 2004, 31–43.

[61]See, e.g., note 60.

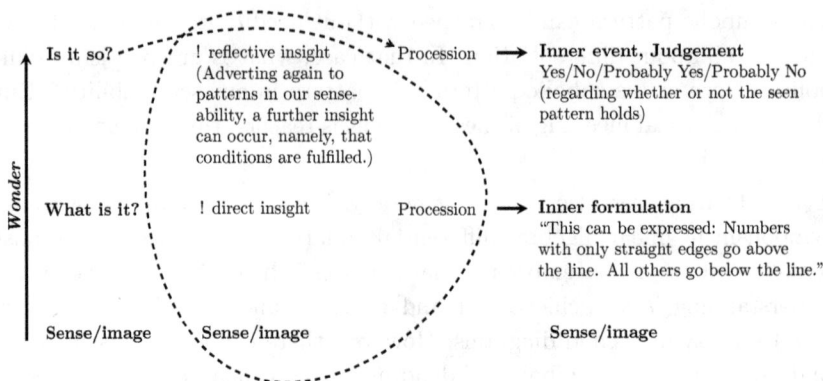

Figure A.4   Reflective inquiry and reflective insight in elementary sequence puzzle A.2.1

to our directly grasped (possible) pattern of mutually defined terms, and operations.

A.4   Descriptive definition and explanatory definition

Let's revisit the 'power rule' sequence: "$y = x$, $y' = 1$; $y = x^2$, $y' = 2x$; $y = x^3$, $y' = 3x^2$; and so on."

In Section A.3, we invited you to observe that in mathematics we enjoy at least two main "genera" of direct insight namely, descriptive and explanatory. Through both, we reach a possible resolution of a 'What is it?' inquiry-poise. But for the power rule, there is only sequence. How do these two resolutions compare? What are we getting at with two answers regarding one sequence?

All along, here, it is a matter of describing what we are doing, of "catching oneself in the act of doing mathematics." From descriptive insight, one understands how to continue the sequence of symbols $y = x$, $y' = 1$; $y = x^2$, $y' = 2x$; $y = x^3$, $y' = 3x^2$. But what do the symbols mean? As in Section A.3, a different understanding is possible. For the power rule sequence, one may also reach a basic understanding of "derivative." [62]

---

[62] In calculus, there are two key *basic* insights, *basic* in the sense of prior to axiomatics. A satisfactory definition of *limit* did not emerge for several decades after the initial discoveries of calculus. But both Newton and Leibniz each had those two key insights

By descriptive insight, then, we understand how to use symbols and diagrams. This makes it possible to "define" symbols and diagrams in the same that way words are "defined" in a typical dictionary. In other words, thanks to descriptive understanding, we can "define" symbols and other mathematical words and diagrams by indicating how to use them. Naturally enough, we call this *nominal definition*.

By explanatory insight, however, we go on to say what the mathematical symbols, words and diagrams mean. So, there is also *explanatory definition*.

We end this section with one more example. To keep the Supplement short, we leave each sentence as an exercise: Defining an ellipse as a "perfectly symmetrical oval" is a nominal definition. Explanatory definition can be reached from the clue that an ellipse is "a circle with two centers." (It will help to make a diagram for drawing an ellipse. This can include an imagined string, two imagined pins and one imagined pencil or pen.) The key insight needed can lead one to (explanatorily) define an *ellipse* to be the locus of all coplanar points satisfying the equation $F_1P + F_2P = constant$. If one has yet not had an insight into "one's image of a two-center circle," then one's "definition" $F_1P + F_2P = constant$ is nominal. Note, however, that in that case nominal definition is at some remove from an imagined "perfectly symmetrical oval" named 'ellipse.' For, in that case, nominal understanding is not of how to use 'ellipse' as a name for an imagined shape but of how to use the name 'ellipse' in conjunction with a (combination of) symbols '$F_1P + F_2P = constant$.'

A.5   Proofs

My beginnings in calculus, by Terrance Quinn

In September of 1980, in the first weeks of my first-year calculus course at the University of Toronto, the lecturer[63] wrote a theorem on the blackboard about continuity of a function $y = f(x)$. He also provided a proof that went

---

(and, of course, much more). Once shared with the scientific communities, those two key insights soon made possible the solution of problems from antiquity and also opened up vast ranges of new lines of inquiry and development in mathematical sciences and engineering. See Terrance J. Quinn, "Getting Started in Calculus," *Problems, Resources and Issues in Mathematics Undergraduate Studies* (PRIMUS), vol. 13, issue 1 (March 2003): 55–74.

[63]I am referring to Prof. Edward Bierstone who, as I understand it, is still at the University of Toronto. See https://www.math.toronto.edu/bierston/. Accessed July 17, 2019.

something like this: "Suppose that $\epsilon > 0$ and let $\delta = \dfrac{\epsilon}{4}$. $\cdots$" He went on to fill two boards with inequalities and ended with words similar to: "$\cdots$ from which we can conclude that $|f(x) - f(y)| < \epsilon$. Since $\epsilon > 0$ was arbitrary, the result follows. ∎"[64]

I trusted our excellent teacher and had no doubt that "the result *did* follow." But during that lecture I also knew that I didn't yet see why. That evening, I started in on trying to unpack the proof. I found some of the inequalities within the proof ("steps" in the proof) to be a bit tricky. Before too long, however, I could follow the proof "step by step," inequality by inequality. I could see that, yes, each step of the proof worked. But I remember that it was also obvious to me that I still didn't really have it. What I mean is, I was aware that even though I could check all of the steps, I didn't see how I might have produced a similar proof myself. Up to that point, in my mathematical education, I had been naively thinking that I already had a good understanding of calculus. In some respects, I did. However, the first weeks of that first semester of calculus were a shock to me and — as I learned — for some (not all) of my classmates. Eventually, along with other "survivors," I started to find my way. For me, the "epsilon-delta business" crystallized in a problem that involved a damped sine function defined over the real line. In hindsight, I see now that my insight in that case was similar to Archimedes' insight regarding the remainder term of a series. Problem by problem and insight by insight, I crawled my way into modern axiomatics for calculus, a.k.a. "elementary real analysis."[65]

In that first-year class, a common type of problem was of the form: "Prove (or disprove) X (where X was some statement)." How to begin? I soon learned that, for me, unless it was "asily done"(in other words, unless I already understood), it was often best to *not* start with the general statement. It usually worked better for me — that is, I worked better — by starting with examples. Of course, I kept the X in mind and oscillated back and forth between examples and thinking about the X statement. But inevitably, the break for me would come while working on an example. That started to take some the mystery out of "proof writing." In other words, "being able to write a proof is not separate from understanding."

I noticed that the order in which the write-up proceeded usually turned out

---

[64]The textbook in the 1980–81 academic year was the edition then available of Michael Spivak, *Calculus*. Cambridge: Cambridge University Press.

[65]Historically, axiomatics for calculus were obtained after several decades of collaboration following the initial breakthroughs of Newton and Leibniz.

to be, to some extent, in a kind of reverse order to what I might call "the order of inquiry and discovery." I now realize that was no accident. I have been recounting some of my early struggles in a challenging and exciting first-year calculus course. But to begin homing in on key and core issues, it is better to now shift to a more elementary example, essential features of which are already described by another author.

Understanding and syllogism

In a circle of, say, unit radius, two diameters, perpendicular to each other, are drawn. [You will need to make a diagram.] From an arbitrary point P on the circumference two perpendiculars PR and PS are drawn to the two diameters. The problem is, What is the ratio of RS to the radius? ... Joining R and S will be an evident thing to do; but it may take a pedagogue to adequately dispose the phantasm by the drawing of another line. The line to draw is the line joining the center to the point P, say OP. Eureka! With the insight there emerges the solution, the relation between RS and the radius.

Note now that the solution can be formulated or thrown into syllogistic form, and this will help you get some light on features of the syllogism that are often misrepresented. We have, therefore, the syllogism:

$$RS = \text{OP}$$

and $\qquad$ OP = Radius;

therefore $\qquad$ RS = Radius.

In this light we may note important characteristics of the procedure. We started not with two premises, but with the conclusion in the form

RS ? Radius.

Our search, through diagram, was for a middle term, and the middle term was supplied as soon as one adverted to the significance of OP. Only then can the syllogism be constructed. To coin an expression for this constructing, one might say that the insight is crystallized into a syllogism. This does not mean of, of course, that somehow insight has been pinned down on a page. What has happened is that we have given the insight symbolic expression. Giving all the relevant insights symbolic expression is by no means always an easy thing to do, even when it can be done. Modern geometers have found fault with Euclid in this matter. There are insights in the *Elements* which are not explicitly acknowledged either in axioms or in the theorems, yet which were not uncrystallizable.[66]

Let me note further that our simple puzzle and solution is a paradigm of how Euclid and company may well have proceeded. They did not proceed step by step down the page of a modern textbook, from the stated theorem to the fully constructed diagram, to the step by step deduction beneath.[67]

We may return now to a brief consideration of the simple symbolic expression of understanding which is the (mathematical) syllogism.

[In mathematics, the syllogism is a] help toward understanding.

The syllogism is not some mysterious replacement for understanding.

One may look at the syllogism as a proof of the conclusion, but this can only mean that the structure facilitates a grasp of the implication of the conclusion in the premises.

One might note, too, that such structures facilitate the checking, the Is-question, relating to that grasp.[68]

---

[66] It concerns drawing a line across the interior of a triangle. One identifies the point where the line crosses the opposite side. Why is there a point of intersection? Felix Klein discusses the problem: "Of course, no one would doubt that, intuitively, but in the framework of axiomatic deduction we need a special axiom, the so-called betweenness axiom for the plane. This axiom states that if a line enters a triangle through a side, it must leave it through the other side — a trivial fact of our space perception which requires emphasis as such, because it is independent of the other axioms. ...If we omit this, as Euclid does, we cannot reach the ideal of a pure logical control of geometry. We must continually refer to the figure" (Felix Klein, *Elementary Mathematics from an Advanced Standpoint: Geometry* (Mineola, NY: Dover Publications, 2004), 88. This is an unabridged republication of an earlier Dover reprint (1949) of the translation first published by The Macmillan Company, New York, in 1939). As is well known, there are non-Euclidean geometries for which the axiom does not hold. For further discussion, see McShane, *Wealth of Self*, 68-9.

[67] "One should note, too, the significance of this for the teaching of geometry. Too often pupils [are asked to] begin, not with the thrill of a puzzle but with the top of the page, and at most get a vague line-by-line comprehension of the theorem. Memory is burdened, and examinations consists in filling out theorems - from the bottom and top of the page! — and passing over the riders" (McShane, *Wealth of Self*, 69). This provides foreshadowing to Part B.

[68] Philip McShane, *Wealth of Self*, 67-9.

A.6 Algebra from arithmetic, "and so on": that is, sequences of "higher viewpoints"

When were children, we learned to count.

"Five baskets of apples and 7 baskets of apples is 12 baskets of apples."

Before long, however, through insight, we learned to 'abstract.'[69]

For instance, in this example, one might go on to understand that whatever we happen to be counting, we can count "how many":

$7 \cdot 3\text{'s} + 5 \cdot 3\text{'s} = (7+5) \cdot 3\text{'s} = 12 \cdot 3\text{'s}$; and

$5 \cdot 9\text{'s} + 8 \cdot 9\text{'s} = (5+8) \cdot 9\text{'s} = 13 \cdot 9\text{'s}$.

Abstracting further,

$7 \cdot 3 + 5 \cdot 3 = (7+5) \cdot 3 = 12 \cdot 3$; and

$5 \cdot 9 + 8 \cdot 9 = (5+8) \cdot 9 = 13 \cdot 9$;

and so on.

And so on?

By insight we are beginning to transition "beyond" arithmetic.

One can solve many arithmetic problems without breaking into, or up into, algebra. That is both historical and biographical. In moving into algebra, however, we do not lose arithmetic. We still understand, for instance, that $5 + 7 = 12$. But now we can also write and understand that $5z + 7z = (5+7)z = 12z$, whatever the number z happens to be.

By insight, one understands and holds together an indefinite range of instances in arithmetic.

Moving further into the new context:

$5 \cdot 3 + 7 \cdot 3 = (5+7) \cdot 3$

$7 \cdot 3 + 5 \cdot 3 = (7+5) \cdot 3 = 12 \cdot 3$

$5 \cdot 9 + 8 \cdot 9 = (5+8) \cdot 9 = 13 \cdot 9$

$\cdots$

?

(direct insight)

from which one can write: $x \cdot z + y \cdot z = (x+y) \cdot z$.

---

[69]Here, 'abstract' is a transitive verb for what we do in thought: "*abstracten*" to draw away, remove, derivative of *abstract* (or borrowed directly from Latin *abstractus*)" `https://www.merriam-webster.com/dictionary/abstract`.

As discussed in Section A.3, that insight may be nominal, as when a student grasps "a pattern in symbolism as symbolism." But one might also understand that x, y and z represent any possible numbers, and that the expression $x \cdot z + y \cdot z = (x + y) \cdot z$ represents a pattern of operations. In that case, we not only know how to use the symbols x, y and z, we also have gotten hold of a correlation of mutually defined terms, and operations, by which one can explain what the expression means. Moreover, with that understanding, we can now just as easily write $a \cdot c + b \cdot c = (a + b) \cdot c$, $U \cdot V + W \cdot V = (U + W) \cdot V$, (U times V) plus (W times V) = (U plus W) times V, and so on.

Transitioning from arithmetic to algebra is sometimes called transitioning to a "higher viewpoint." In the case of arithmetic and algebra, both "viewpoints" are *systems*. The word 'higher' is a spatial metaphor, but it serves nicely. For, as revealed in instances, by understanding in algebra one has control over indefinite ranges of arithmetic computations.

A key point: Self-notice that images by which we reach the higher viewpoint of algebra are diagrams and symbols from lower viewpoints such as arithmetic and elementary geometry.

We are not asserting this key point as a conclusion from a "theoretical framework" or model for mathematical development. It is an observation reached by adverting to instances to one's own experience.[70]

In mathematics, does transitioning from lower to higher viewpoints happen often?

In order to make beginnings toward being able to answer this Is-question one would need to have (considerable) experience in mathematics. One would also need to attend to what one has been doing. The question calls for and depends on growth in mathematics as well as growth in the kind of "balanced" attention being invited in this Supplement. In other words, there is the further challenge of adverting to, noticing and discerning as precisely as possible what one is doing when one is doing mathematics.[71]

---

[70] As you might already be seeing, this has implications for teaching. See Part B.
[71] The task is further described in Section A.8.

A.7 Correlations, concepts and other fruit of understanding

Correlations

In coordinate (Cartesian) geometry, the "parent parabola" is defined by the equation $y = x^2$. Other parabolas are obtained by adjusting the scale of one or both of the axes, by translation, and by rotation.

We start with this definition because it will be familiar to many teachers. It is what is often given in high school and first year undergraduate books.

The present challenge is to make a beginning in "unpacking" (what turn out to be "layerings" of) insights that often are neither expressed nor adverted to.

Let's begin with our focus on $y = x^2$, an algebraic formula. In that context, what are the x and the y? By using a pair of symbols, we refer to an indefinite range of pairs of numbers, some of which are (0, 0), (-1, 1), (1, 1), (-2, 4), (2, 4), (-3, 9), (3, 9), and so on. In each pair, what is x? It represents every number that squares to give a corresponding number y. What is y? It represents every number that is the square of some x. The x and the y are mutually defined by a correlation that can be expressed by '$y = x^2$'. Note that in this context, terms, operations and equality are algebraic. But in that correlating, we also understand and correlate an indefinite range of pairs of numbers where in each case, terms, operations and equality are arithmetic.

That's a start. However, we haven't finished "unpacking." Why not? Coordinate geometry is not merely algebra. In coordinate geometry, the x and the y are coordinates. In other words, the x and y refer to lengths along axes that, by hypothesis, have scales. So, let's look more closely to our underlying (or perhaps latent) understanding of an x-axis.

One can imagine a distinguished straight line in the Euclidean plane[72] and call it the 'x-axis.' In coordinate geometry, we also have a scale along that distinguished line (and, indeed, along all lines in the same Euclidean plane).

---

[72]What is a *straight line* in Euclidean geometry? (Here, we avoid the extensive literature regarding "primitive concepts," "primitive," and "vicious circles in logic." Such problems need to be handled, but not here.) In Book I of the *Elements*, Euclid provides what evidently are nominal definitions: "Def. 1.2. *A line is a breadthless length.*" "Def. 1.4. *A straight line lies equally with respect to the points on itself.*" Explanatory definition is obtained when a straight angle is defined to be the sum of two 90-degree angles. In that way, we go beyond merely describing what we imagine. In that case, a straight angle is defined by correlating terms in the system.

And so we can write $x = 0$, $x = 1$, $x = 2$, $x = 22/7$, $x = e$, $x = \pi$, and so on. In coordinate geometry, these numbers refer to lengths along the distinguished axis.

This may seem obvious enough. But is it?

Let's take one of these, $x = 2$, say. In this context, to what do we refer with the number 2?

Here, $x = 2$ is a "distance" along the x-axis.

How is that distance determined?

You might remember that the task here is not to invent something new but to discern elements in what we already do. Distance is given in terms of a *unit distance* along the x-axis.

OK, but what is a *unit distance*?

In physical measurement, a *unit distance* is whatever happens to be convenient. This might be the length of one's thumb (an inch, or in French, 'un pouce,' which means 'thumb'), a handy length of wood, the distance from the tip of one's middle finger to one's elbow (cubit), a "standardized cubit" such as the length of a piece copper used by the Sumerians in the third millennium B.C.E., and so on. In modern times, a 'meter' is the length of a bar of platinum held in Paris, France.

In mathematics, distance is not a physical length. Still, there is nothing to stop us from imagining and thinking about a unit length,

" _____ ".

Physical lengths can and do change. Two people with different arm lengths will yield different cubits. A piece of wood may dry out and its length may then change. Or again, suppose we have two metal rulers with "cms" etched along their sides. In other words, "1 cm" is the unit length. If one of the rulers is used at different locations on a hot wood stove then "1 cm" at one location can be longer than "1 cm" at another, when compared with the second ruler that is kept from direct contact with the hot stove. But if there is no fire in the stove, and if no finer measurements are needed then, for practical purposes, no problems arise. That is, the unit length is "invariant." hrough the ages, builders, carpenters, architects and engineers have relied on such invariance when designing and building structures — from wooden shacks to pharaohs' pyramids, from modern homes to city skyscrapers.

In classical geometry, we "ake this idea and run with it." We suppose that an imagined unit length is invariant in the sense that an imagined unit length at one location is the same as an imagined unit length at another.

Let's now return our focus to the x-axis. Let O_x be a (distinguished) point on the x-axis. The point O_x is called "origin." Let X be another point on the same axis, a point that is, say, 'two units' "to the right" of O_x. Notice our ongoing reliance on imagination, e.g., "to the right," and "to the left."

We have an imagined x-axis which is an imagined distinguished line in the Euclidean plane. By insight, we suppose that the line extends indefinitely in both directions:

"... ——————————————————————— ..."

We imagine two points on this imagined x-axis, O_x and X:

$$O_x \qquad X$$
... ———————•————————•——————— ...

In the imagined length between these two imagined locations, one can imagine fitting copies of the imagined unit length, " —— " and " —— ", distinct line segments that, by hypothesis, are otherwise identical. In particular, we suppose that they have the same length.

A handy diagram for this is:

$$O_x \qquad X$$
•————————•

——— ———

So, in our understanding, we correlate an imagined line segment with copies of an imagined unit length:

———————

——— ———

But we also correlate an imagined " ———————— " with the number '2', a diagram for which is:

———————

2

In coordinate geometry, "$x = 2$" means that we take both of these correlations "together," a diagram for which is:

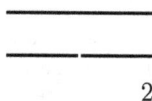

———————

———————

2

In other words, "together" means that we need a further insight by which we obtain, grasp, or reach a further correlation. That is, we correlate two correlations.

What we find then is that, in coordinate geometry, asserting that $x = 2$ is an expression of an insight by which one holds together a "layering" of insights: one correlates two correlations. Similarly, $x = 3$ is an expression for correlating two correlations; and so on for all values of x; and similarly for all values of y.

We started with the expression $y = x^2$. If we only focus on numbers, then our understanding is an algebraic correlation. In coordinate geometry, however, the x's and y's are lengths along two distinguished (and perpendicular) lines in the Euclidean plane, a plane in which there is, by hypothesis, an invariant unit length. Unpacking key insights, the x-values represent an aggregate of correlations of correlations; as do the y-values. In *coordinate geometry*, then, $y = x^2$ is an expression of a *correlation of correlations of correlations*.

Concepts

In Euclidean geometry, we speak of 'points,' 'lines,' 'line segments,' 'length,' 'invariance,' and so on, as well as the entire 'Euclidean plane.'

What are these?

What, for instance, is a *point*?

We can imagine a small dot on paper (or, say, on a computer screen). An imagined dot has breadth and so we can imagine two smaller dots within an imagined dot. Still, "the point" is "clear," is it not? That is, attempting to "pin down" a location unambiguously, we need to keep making an imagined dot smaller. So long as a dot has any breadth or depth, further dots can be imagined within the imagined dot.

By **insight**, then, we define a *point* on the plane to be "a location that has no breadth and no depth."

Or, in translation from Book I of Euclid's *Elements:* *"A point is that which has no part."*

A *point* then, (**self-**) **evidently** is the fruit of insight! It is also a concept.[73]

---

[73] "concept: noun, (Entry 1 of 2) 1: something conceived in the mind: thought, notion, 2: an abstract or generic idea generalized from particular instances" https://www.merriam-webster.com/dictionary/concept.

In a similar way, that is, by adverting to experience, one can find that lines, planes, invariant length, and so on, are also the fruit of insight and also are concepts.

Continuing in this way, what becomes (self-) evident is that while it is true that images and patterns in our sense-ability on which we focus inquiry can be said to be "primitive," points, lines and other concepts are the fruit of understanding and, in particular, are neither reducible to imagination nor are they primitive elements in mathematics.

A.8   The historical context for teachers (and scholars)

How do we go on from elementary and preliminary exercises? It is for each teacher and scholar in mathematics to work out how far you need to go. "Just as in any subject, one masters the essentials by varying the incidentals."[74] In so far as one has the time and the proclivity, one can identify key diagrams and symbolic expressions, helpfully directed questions, key insights, skills, and sequences of such in one's own mathematical development.

You might observe that, unless one is breaking new ground, when understanding mathematics one is also sharing in key insights from the historical development of the field. Note that this is not a statement regarding "the historical method" for teaching. This is an *observation* about learning mathematics, for instance, algebra, or calculus. How else is one to know and teach "completing the square" that was discovered in various times and places in antiquity, or "the calculus" that was discovered by Newton and Leibniz, and know these in ways that allow one to teach them effectively, unless one has not (at least partially) identified key questions and insights? At the same time, (self-) evidently, it is always our own inquiry and our own insights that we experience. Our own development is our main source of data[75] on mathematical development.

It was Bernard Lonergan who first brought attention to the need and possibility of the balanced method to which we are inviting mathematics teachers. And so we end Part A of this Supplement with a quotation from Lonergan on historical understanding [of mathematical development].

---

[74]Lonergan, *Insight*, 56.
[75]See note 50.

The original quotation is from a discussion of historical understanding of a discipline.

The history of [mathematics] is in fact the history of its development. But this development, which would be the theme of a history [of mathematics], is not something simple and straightforward but something that has occurred in a long series of various stages, errors, detours, and corrections. To the extent that the one studying this movement learns about this developmental process, one already possesses within oneself an instance of that development which took place perhaps over several centuries. This can happen only if one understands both the subject and the way in which he or she learned about it. Only then will one understand which elements in the historical developmental process had to be understood before others, which were the causes of progress in understanding and which held it back, which elements really belong to that particular science and which do not, and which elements contained errors. Only then will one be able to tell at what point in the history of the subject there emerged new visions of the whole and the first true system occurred, and when transitions took place from an earlier to a later systematic ordering; which systematization was simply an expansion of the former and which was radically new; what progressive transformations the whole subject matter underwent; how everything that was previously explained by the old systematization is now also explained by the new one, as well as many other things that the old one did not explain [as in the discoveries in mathematics, for example, by Euclid, Archimedes, Apollonius, Pappus, Al-Khwarizmi, Descartes, Fermat, Newton, Leibniz, Cauchy, Fourier, Riemann, Galois, the great analysts, and others]. Only then will one be able to understand what factors favored progress, what hindered it, why, and so forth.

Clearly, therefore, [master teachers of mathematics] have to have a thorough knowledge and understanding of the whole subject. And it is not enough that they understand it in any way at all, but they must have a systematic understanding of it. For the precept, when applied to [the history of mathematics], means that they must understand successive systems that have progressively developed over time. This systematic understanding of a development ought to make use of an analogy with the development that takes place in the mind of [a teacher] who is learning the subject, and this interior development within the mind of the [teacher] ought to parallel the historical process by which [mathematics] itself developed.[76]

---

[76]Bernard Lonergan, Early *Works on Theological Method 2*, vol. 23 in the *Collected Works of Bernard Lonergan*, translated by Michael G. Shields, edited by Robert M. Doran and H. Daniel Monsour (Toronto: University of Toronto Press, 2013): 175–177. For a somewhat dated but still excellent source on mathematical development, see, Eric Temple Bell, *The Development of Mathematics*. New York: McGraw-Hill, 1940. Second Edition: New York, McGraw-Hill, 1945. Reprint: Dover Publications, 1992. University of Toronto Press, 2013): 175–177. For a somewhat dated but still excellent source on mathematical development, see, Eric Temple Bell, *The Development of Mathematics*. (New York: McGraw-Hill, 1940. Second Edition: New York, McGraw-Hill, 1945. Reprint: Dover Publications, 1992).

A Few Implications for Teaching

It is for brevity that we merely state a few results. To bring these out pedagogically would need a few chapters. However, we hope that providing these few results will serve as a guide and invitation to further reflection. If you have worked through the exercises in Part A (or similar ones), you may get to some (or all) of these yourself. As pointed to in Section A.8, further details and implications will be discovered through one's ongoing growth in mathematics and in self-attention in mathematics.[77] We do not comment on curricula of any departments or colleges involved in the ongoing challenging and creative work of meeting students' needs and program needs in diverse circumstances.

Part of the invitation is for a teacher to make progress in identifying and distinguishing nominal and explanatory understandings, in one's own understanding.

As experience reveals, nominal understanding includes understanding by which one reaches competence with symbolic techniques and computations.

The task of pedagogy invites teachers to know as much as possible about where, in precise terms, the content of a course fits in the historical development of the field. See Section A.8.

By the same token, in order to be able to help a student make progress in the field, it will help if a teacher learns as much as possible about where a student's understanding is in relation to the historical development of the field.

As expressed by Figure A.1, by adverting to one's experience in doing mathematics, it becomes evident that understanding emerges through inquiry

---

[77]See note 45. "balanced method."

"into" images in one's sense-ability. Consequently, in a given mathematical context, it will help if a teacher is familiar with diagrams and symbolisms that help raise and direct student inquiry toward reaching key insights appropriate to a context.

To help a student reach a higher viewpoint, it will help if a teacher can provide diagrams and symbolisms from lower viewpoints by which a student (or group of students) can reach key insights needed to begin breaking into the higher viewpoint. See Section A.6.

For a century and more, there has been ongoing discussion about whether or not teachers should "use history" to teach mathematics. An advantage of the "historical method" is that it has the potential for helping students raise questions and reach key insights in the field. But of course not all of the searchings from history can be included. A selection needs to be made and tailored to an audience and to a curriculum. In that sense, a teacher needs to "make history better than it was." More precisely, a teacher can make progress in identifying specific developmental sequences.[78] In that way, for instance, in two leisurely 45 minute classroom sessions one can help senior high school mathematics students and freshmen level general calculus students reach the two key insights had by both Newton and Leibniz from which the entire body of calculus is developed.[79]

Pedagogy called engaged learning has been a fruitful idea. The importance and relevance of modern psychology and neuroscience (so, e.g., behaviour, environmental considerations, learning style, group work, and so on) to pedagogy cannot be denied. But might we not also ask, engaged in what? We are back at the need and possibility of identifying helpfully directed diagrams and symbols by which to subtly guide student inquiry toward specific and key mathematical insights.

Only a small percentage of students go on to become mathematics majors who, in addition to needing *basic*[80] key insights, also go on to (sequences of) axiomatic systems. But students in applied mathematics, engineering, applications and general service courses also need to be helped so that inquiry be directed toward the emergence of initial or basic key insights. Why? Without basic insights (which, as experience reveals, precede the emergence of axiomatic system), understanding is mainly nominal. Even if a student is only wanting to be able to do routine computations in a

---

[78]See note 76.
[79]See note 62.
[80]See note 62.

practical career, if understanding is merely nominal, they will be unable to proceed whenever boundary conditions do not sufficiently mesh with what is only nominally familiar.

Modern textbooks speak of "understanding concepts" and "conceptual understanding."[81] It is said (or assumed) that "concepts are primitive elements" and that mathematical understanding is a matter of "connecting concepts" (that one has prior to understanding). It is also now popular to assert analogies between structures of computer programs and human understanding. There is an 'Is it so?' question here. By adverting to instances in our experience (see, e.g., Section A.7), it is (self-) evident that while there are "primitive elements" in mathematics (for instance, in Euclidean geometry), it is not concepts that are primitive. And the experiential basis of alleged analogies with computer programming are not found in mathematical understanding. They are obtained, rather, by correlating structures of computer programs with structural features of hypothetical conceptual models of mathematical understanding. On the other hand, adverting to experience in mathematical understanding reveals that points, lines, invariant unit length, and other familiar concepts emerge from insight. In other words, concepts are the fruit of understanding.

Relative to arithmetic, algebra is a higher viewpoint. As already described, one reaches a higher viewpoint by abstracting[82] from instances in a lower viewpoint.

This sheds light for us on the problem of giving students calculators and other computational technologies too soon. If a student does not have a good understanding and technical competence in arithmetic, then they are lacking most if not all of the symbolic data, diagrams and understanding needed in order to break through to the higher viewpoint of algebra. Similarly, if a student lacks nominal and explanatory understanding in algebra and coordinate geometry, they will be lacking most if not all experience needed in order to break through to the higher viewpoint of calculus. Again, there are transitions from algebra to abstract algebra, from calculus to function theory, and so on.

Evidently, building up an axiomatic system relies on understanding that is beyond initial or basic understanding (see, e.g., Section A.5). This helps reveal a common misdirect, that is, when a chapter or lesson begins with "preliminary concepts," "axioms" and/or "axiomatically correct

---

[81]See Part C.
[82]See note 69.

definitions." Certainly, that approach is dull and is a way to undermine engaged learning. Attempting to start with axioms and preliminary concepts is, in fact, attempting to start with answers to questions from a further context. The approach does not bring the student into an inquiry zone that promotes emergence of initial key insights. But this is another face of the misconception in mathematics education that is often called "conceptual understanding."[83]

Whether implicit or explicit, a teacher's view of mathematical development influences how and what we teach. Whatever one's context in teaching, and whether private or shared with colleagues, there is the possibility of reflecting on one's own experience in mathematics and on what one hopes one's students also will learn. In other words, as invited in Part A, there is the possibility of growing in being able to advert to and precisely distinguish elements in one's own experience. If one is already a successful teacher, the invitation is to accurately identify sources of one's success and that way become an even more effective teacher. If one is struggling as a junior teacher, the invitation can include the task of identifying gaps in one's mathematical understanding, inconsistencies in what one understands, and elements of one's own understanding (specific combinations of nominal and explanatory, in and across diverse contexts), and to make efforts to bring one's learning and teaching into better harmony with the historical field and the needs of one's students.

---

[83] See three paragraphs above. On the other hand, 'conceptual understanding' is just a name. Many excellent teachers use the term in a positive sense, namely, when referring to student understanding that goes beyond mere technical competence. Implicitly, they are referring to both nominal and explanatory understanding in mathematics. The name "conceptual understanding" however is, in fact, inherited from the philosophical literature and in that context refers to what does not occur in mathematics. That is, as experience reveals, we do not understand concepts but, rather, concepts emerge from understanding.

PART C

Observations Regarding Modern Mathematics Education

For decades, emphases in mathematics education have been on "theoretical frameworks."[84] Attempts to justify a theoretical framework are made by working out features of, and in, a framework; designing lessons based on such derivations; and then appealing to observations and statistical analyses of, for instance, student scores, student behavior, reported feelings, and numerous other "aggregate-events" remote to source events that are mathematical inquiry and understanding itself.[85] Unfortunately, this approach has been promoting fundamentally mistaken views of mathematical understanding.

We provide only a few remarks. We are not attempting to give a scholarly discussion of contemporary mathematics education.[86] Our strategy here mainly is to tease at a few strands of the problem. Among other things, this will help teachers begin see that it is possible to personally assess models that one may be asked to implement.

Let's start by looking to the Introduction of the well-known and influential paper by Dubinsky and McDonald (2001).[87]

---

[84]Schoenfeld, 2016. For a survey of "theoretical frameworks used in the field of Calculus education," see David Bressoud, I. Ghedamsi, V. Martinez-Luaces, G. Törner, G. *Teaching and Learning of Calculus*, ICME-13 Topical Surveys, Springer Open (2016) 1-37. https://www.springer.com/gp/book/9783319329741.

[85]See, e.g., Schoenfeld, 2016.

[86]That will be a community achievement, coming from a new kind of collaboration on a global scale.

[87]E. Dubinsky and M. A. McDonald, "APOS: A constructivist theory of learning in undergraduate mathematics education research," in D. Holton et al. (Eds.), *The Teaching and learning of Mathematics at University Level: An ICMI Study* (Dordrecht, Netherlands: Kluwer Academic Publishers, 2001): 273–280.

> We do not think that a theory of learning is a statement of
> truth and although it may or may not be an approximation to
> what is really happening when an individual tries to learn one or
> another concept in mathematics, this is not our focus. Rather
> we concentrate on how a theory of learning mathematics can help
> us understand the learning process by providing explanations of
> phenomena that we can observe in students who are trying to
> construct their understandings of mathematical concepts and by
> suggesting directions for pedagogy that can help in this learning
> process.[88]

Note that, in the first sentence, the authors express a lack of concern for
whether or not their theory is "a statement of truth." They then remove
themselves from the task of inquiring into "what is really happening when
an individual tries to learn": "this is not our focus." Their focus is, instead, "phenomena that we can observe in students."[89] The last sentence
of the paragraph reveals that in their inquiry into learning, they presuppose a theory of learning. That this, they presuppose a version of constructivism which, in this context, is to the effect that students "learn one or
another concept in mathematics"; that they "construct their understanding of mathematical concepts"; (later in the paper) that when learning
mathematics one "[perceives] mathematical problem situations"; and that,
when learning, "an individual is developing her or his understanding of a
concept."

If you have done exercises such as those in Part A (or beyond) and have
made beginnings in understanding your own understanding in mathematics,
do not the claims made in the paper cry out for correction?[90]

Perhaps, though, the resulting pedagogy stands. So, let's look at what
Dubinsky and McDonald suggest for teaching *cosets*:

> Pedagogy is then designed to help the students make these
> mental constructions and relate them to the mathematical

[88] Dubinsky and McDonald, 2001.

[89] If one's focus is on "phenomena that we can observe in students" then inquiry will
not be about mathematical insight but, rather, patterns in "phenomena what we can
observe in students" who do not yet understand. Note, too, that if (as in APOS and
its applications) those phenomena are not explanatorily defined, statistical analysis has
little or at most preliminary explanatory significance.

[90] Regarding concepts, see Section A.7. The confusion is historical and not unique to
the work of Dubinsky and McDonald.

concept of *coset*. In our work, we have used cooperative learning and implementing mathematical concepts on the computer
in a programming language which supports many mathematical constructs in a syntax very similar to standard mathematical
notation.

Evens and odds, "clock arithmetic" and other similar arithmetic and geometric groupings are learned by children. In that way, students learn much
of what is needed in order to prepare the way for reaching a higher viewpoint that includes an understanding of cosets. In our own mathematical
development, we learned mathematics, and did not make use of syntax
"similar to standard mathematical notation." Indeed, in the historical development of cosets, founders did not appeal to computer programming
language. Is it not evident that syntax (of computer programming) "very
similar to standard mathematical notation" is, in fact, a distraction from
the *mathematical* problem?

Why did Dubinksy and McDonald advocate that approach to teaching
cosets? Why attempt to use something similar to mathematics to teach
mathematics instead of helping students learn mathematics itself? Indeed,
why might that be relevant (or not) when "what is really happening" is, as
they suggested, not the focus and not known? They explain it as follows:

Their design instruction focused, not directly on mathematics, but on some
model of how the topic in question can be learned.[91]

Dubinsky and McDonald had a remarkable dedication to the cause of mathematics education. As scientists, however, they chose to not focus on "what
is really happening" but rather on "phenomena that we can observe in students." In that approach, they also "took themselves out of the equation"
and so did not avail of experience in mathematical understanding.

Constructivist models allegedly account for mathematical understanding.
And so, to reveal the presence of flaws, all that is needed is a counterex-

---

[91]Ed Dubinsky, "Using a Theory of Learning in College Mathematics Courses,"
MSOR Connections (2000), 1.10.11120/msor.2001.01020010. This article was originally published in Newsletter 12, TaLUM, Teaching and Learning Undergraduate
Mathematics subgroup, 2001. It is available online: http://www.math.wisc.edu/ wil-
son/Courses/Math903/UsingAPOS.pdf. Accessed July 17, 2019. The tradition continues. The literature is extensive. For a point of entry into the literature see, e.g., L.
Benton, C. Hoyles, I. Kalas and R. Noss, "Designing for learning mathematics through
computer programming: A case study of puplis engaging with place value," *International
Journal of Child-Computer Interaction*, vol. 16 (2018), 68–76.

ample. But by doing exercises in Part A (or similar ones), one can obtain numerous counterexamples across a range of instances in elementary mathematics.

Concluding comments for Part C

As already alluded to, conceptual models of human understanding are not new. Part of a larger tradition called "conceptualism," roots of contemporary constructivism go back at least as far as claims made by Duns Scotus (1266–1308).[92]

Aspects of engaged learning models are proving to be significant and are partly grounded in progress in human neuroscience and human psychology. But in instances, where a particular mathematical result is needed in what, *specifically*, is a student to be engaged? And so, a model might be partly valid but also call for development. On the other hand, there are "theoretical frameworks" that are mainly speculative.

How does one tell the difference and so make positive use (or not) of models encountered?

One can always ask:

Does the model explain instances, in detail, in *my* experience?

If a model of mathematical learning does not bear out in one's own instances of mathematical learning, then (self-) evidently the model is in some way flawed. Note that what is novel[93] here and most important is not this or that model. Neither are we suggesting that teachers or other readers believe what we have only very briefly touched on this Supplement. Rather, we are inviting teachers to a way by which one can check for oneself and so obtain "ground to stand on," one's own experience. We have found, and for instance, we hope that you are finding, that by adverting to one's own experience (by doing exercises like those given in Part A, and more), one can make progress in identifying and distinguishing orderings of key and core

---

[92]In Scotus' speculative theory of knowing, "[i]ntellectual intuitive cognition does not require phantasms; the cognized object somehow just causes the intellectual act by which its existence is made present to the intellect. As Robert Pasnau notes, intellectual intuitive cognition is in effect a 'form of extra-sensory perception' (Pasnau [2002])" (Williams, Thomas, "John Duns Scotus," *The Stanford Encyclopedia of Philosophy* (Spring 2016 Edition), Edward N. Zalta (ed.), https://plato.stanford.edu/archives/spr2016/entries/duns-scotus/. See https://plato.stanford.edu/entries/duns-scotus/#MatForBodSou.

[93]The method is not new but will be new in mathematics education. See note 43.

events in sequences of instances in one's own mathematical development.

Among those who engage in such exercises, there will be differences in mathematical background and in descriptions of elements in one's inquiry and understanding.[94] So, we can expect that there will also be differences in views regarding learning and teaching mathematics. Like in any serious science, such differences will not be resolved by logical debate about models. However, to the best of our ability we can attempt to spell out and discuss aspects of our own experience, inquiry and growth in mathematics.

---

[94]Although, the basic structure given in Figure A.1 is invariant. If one claims that a different structure accounts for one's understanding, is it because one understands something about an understanding that one does not have; or that one judges what one does not understand? In the terminology of philosophy of science, there results what is called 'performance-contradiction.' Philosophical terminology is not the main issue here. As in the text above, a fundamental question always is: "Does a model explain *instances*, in detail, in *my* experience?" Is this *me*?

Index